DCC
PROJECTS & APPLICATIONS
VOL. 4

Kalmbach
Media

About the author

Dr. Larry Puckett has been *Model Railroader's* DCC Corner columnist since 2015. He is the author of two previous Kalmbach books: *Wiring Your Model Railroad* (2015) and *Wiring Projects For Your Model Railroad* (2018). Larry has also written more than 200 articles for various hobby publications including *Model Railroader, Model Railroading,* and *Railroad Model Craftsman*. A long-time modeler, Larry is a former research scientist who retired after a 33-year career as a water-quality specialist with the U.S. Geological Survey.

The material in this book has previously appeared in *Model Railroader* magazine. Unless otherwise noted, all text and photos are by the author. The text has been edited to reflect current manufacturer products and information as much as possible, but may occasionally mention products that are no longer available.

Kalmbach Media
21027 Crossroads Circle
Waukesha, Wisconsin 53186
www.KalmbachHobbyStore.com

© 2019 Kalmbach Books
All rights reserved. This book may not be reproduced in part or in whole by any means whether electronic or otherwise without written permission of the publisher except for brief excerpts for review.

Published in 2019
28 27 26 25 24   3 4 5 6 7

Manufactured in China

ISBN: 978-1-62700-688-0
EISBN: 978-1-62700-689-7

Editor: Jeff Wilson
Designer: Lisa Bergman

Library of Congress Control Number: 2011499032

Contents

Introduction 4

1 Digital Command Control basics
Switching to DCC 8
Selecting the right DCC system 10

2 Layout wiring and power supplies
Layout wiring for DCC 12
Power booster basics 14
Dividing layouts into power districts 16
DCC power supplies 22
Improving power and signal reliability 24
Reverse loops, turntables, and wyes 26
Avoiding short circuits with DCC 28
Wiring all-live turnouts 30
Wiring power-routing turnouts 32
Electrical troubleshooting 34
Computer interface options for DCC 38
Tips for better soldering 40

3 Decoder and sound installation
Are your locomotives DCC friendly? 42
Installing a basic decoder in a classic model 44
Sound and DCC for an HO Atlas U30B 46
Adding sound to an N scale steam locomotive 49
Replacing a factory diesel decoder 52
Adding sound to an older HO Atlas RS-1 55
Replacing an old HO steam sound decoder 58
Choosing speakers for sound installations 60
Stay-alive modules 62
Installing sound with DCC in a brass diesel 64
Adding a LokSound decoder to an older diesel 66
Dual engines with a WOWSound decoder 68
Packing sound into a vintage HO Kato switcher 70
Back-EMF and a quick decoder installation 72
Squeezing sound into an HO SW1 switcher 74
SoundCar decoder installation 78

4 Programming and layout projects
Use DecoderPro to simplify programming 80
Tips for programming LokSound decoders 82
Getting the most from automatic functions 84
Remapping functions for consistent operation 86
Basic, universal, and advanced consisting 88
Speed matching for DCC consists 90
Turnout control with accessory decoders 92
Add DCC and sound to a turntable 94

Introduction

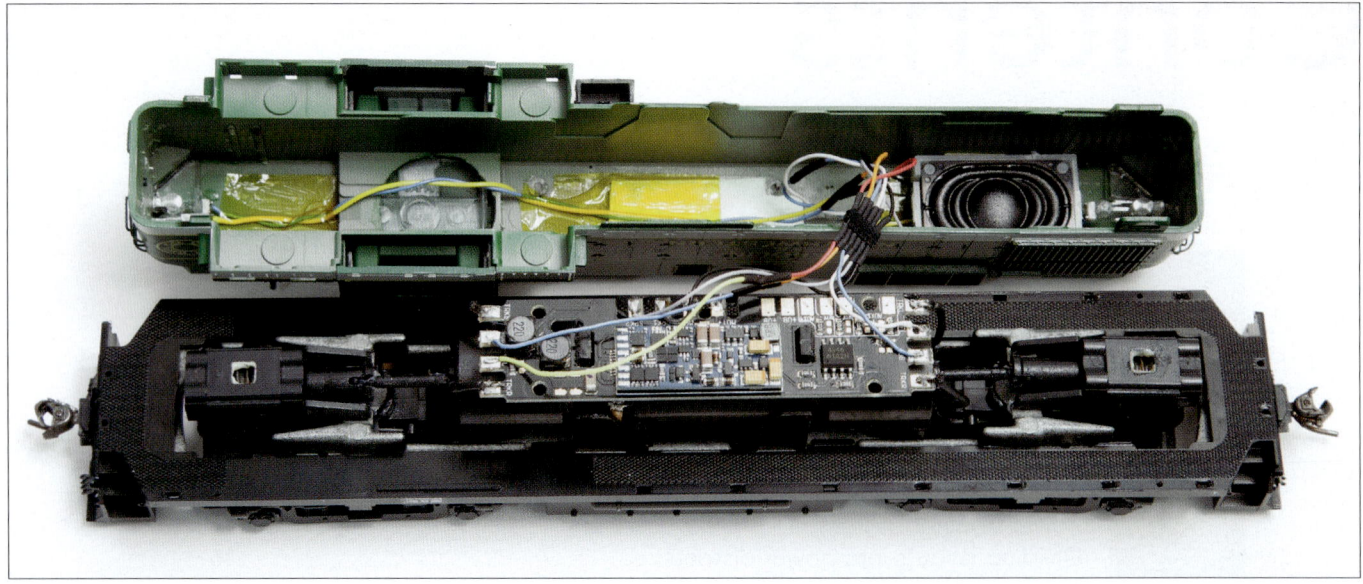

1. Decoder installations are still a key area of interest for modelers. This is an ESU LokSound decoder that Larry Puckett installed in an older HO scale Atlas Alco RS-11. You can see details of the installation on page 66.

Model railroading and Digital Command Control (DCC) are constantly evolving as the pace of technological advancement speeds up and subsequently trickles down to our hobby. In volume 4 of *DCC Projects and Applications* we'll be looking at how some of these advancements have added to the enjoyment of model railroading since Volume 3 was released in 2015.

Most of the articles in Volume 4 were originally published in my DCC Corner column in *Model Railroader*, which I took over in 2015. Some other authors have contributed to this volume; we have noted those items with bylines along with their titles.

To begin, I'll bring you up to date on many of the new advances and products from DCC manufacturers over the last four years—and there are a lot of them. As always there have been major enhancements in decoders, especially where sound is concerned. Manufacturers have also taken advantage of technological advancements to provide new features to command stations, and some new systems have entered the fray. We also have seen the industry borrow from other technologies to introduce new options in WiFi-based wireless throttles, speakers, and devices to help us get over dirty track and dead frogs.

In section 1 we'll review some of the basics of DCC and the advantages it offers us over standard DC power-based control methods. After that we'll go over some of the things to consider when selecting a DCC system.

Wiring a layout for DCC operations is very important, since there are concepts to be aware of that are DCC specific. Section 2 will provide detailed information on selecting the correct size wire, installing power buses, planning power districts, dealing with short circuits, and troubleshooting in general. A special area of interest to many will be the material on differences between power-routing and all-live turnouts and guidelines for using them with DCC.

Although many model locomotives being manufactured today are offered with factory-installed DCC decoders, many model railroaders still prefer to install their own, especially sound decoders, **1**. This has become increasingly important as model manufacturers use decoders from a variety of different companies. In the long run many modelers have found it less confusing if all their locomotives have the same brand decoders. In section 3 we will show how to install a variety of decoders in steam and diesel locomotives. We'll also explore the very important topic of choosing a sound decoder and the right speaker and the correct way to install it.

Programming decoders is a subject that scares many model railroaders but can be very simple, and in section 4 we'll provide some guidance and lessen the potential anxiety. While many never find it necessary to change much more than the decoder address and sound volume, some will want to do more to enhance the enjoyment of their sound-equipped locomotives.

To help with that we'll discuss making up consists of multiple locomotives, changing which buttons control the various functions, and which ones play automatically. We'll also provide some information on using the free computer program DecoderPro to tackle the more complex aspects of sound decoder programming.

Finally in chapter 5 we'll talk a little about using accessory decoders to enhance operations.

What's new in DCC

I want to give my innovation award to the unknown person who decided to see how cell phone and tablet computer speakers would sound in a model locomotive. These tiny devices, which can be as small as your pinky fingernail, **2**, have made it possible to install sound decoders in models I previously never would have attempted. And the

by Larry Puckett

2. So-called "sugar cube" speakers, designed for use in cell phones and tablets, are compact, provide good sound quality, and are easy to fit into many locomotive shells.

amazing thing is they provide excellent sound. My first experience with these came in 2015 when I installed one in an HO Atlas RS-1 (page 55). The 11mm x 15mm speaker fit nicely in its hood and is still functioning perfectly over three years later.

The speakers themselves are only a couple millimeters thick, but require a sound box or enclosure which usually adds about 10mm to the thickness. Streamlined Backshop sells enclosures in a variety of configurations that must be glued to the speaker, while the enclosures from TCS and Tony's Train Exchange provide a drop in fit. Sugar cube speakers have become my go-to choice for diesel installations, but I still prefer a larger high-bass speaker in my steam locomotives.

Stay-alive devices

Although stay-alive devices have been available since about 2005, they have become increasingly popular in the last few years and almost all major decoder manufacturers have added a version to their product line. These basically consist of capacitors, voltage regulators and associated circuitry that provide backup power for decoders. This backup power is usually enough to keep a decoder operating for several seconds.

The popularity of these devices has greatly increased along with that of sound decoders, because they prevent annoying sound interruptions. They are especially useful for preventing locomotive shutdowns and jerky operations due to dirty or uneven track and unpowered turnout frogs. Recently, manufacturers like Bachmann have begun including stay-alives along with factory-installed sound decoders in locomotive models such as the ACS-64 electric and Pennsylvania K4 streamlined Pacific locomotives.

Powered frogs

For many years modelers have used auxiliary electrical switches on turnout switch machines to control the polarity of their turnout frogs. Several manufacturers including Micro Engineering and Peco make turnouts with built-in spring mechanisms that hold turnout points against the stock rails, eliminating much of the need for switch machines.

However, it's still desirable to power the frog and control its polarity. Tam Valley Depot makes circuits called Frog Juicers, 3, designed to automatically detect a short at the frog and correct it in only about 150 microseconds, long

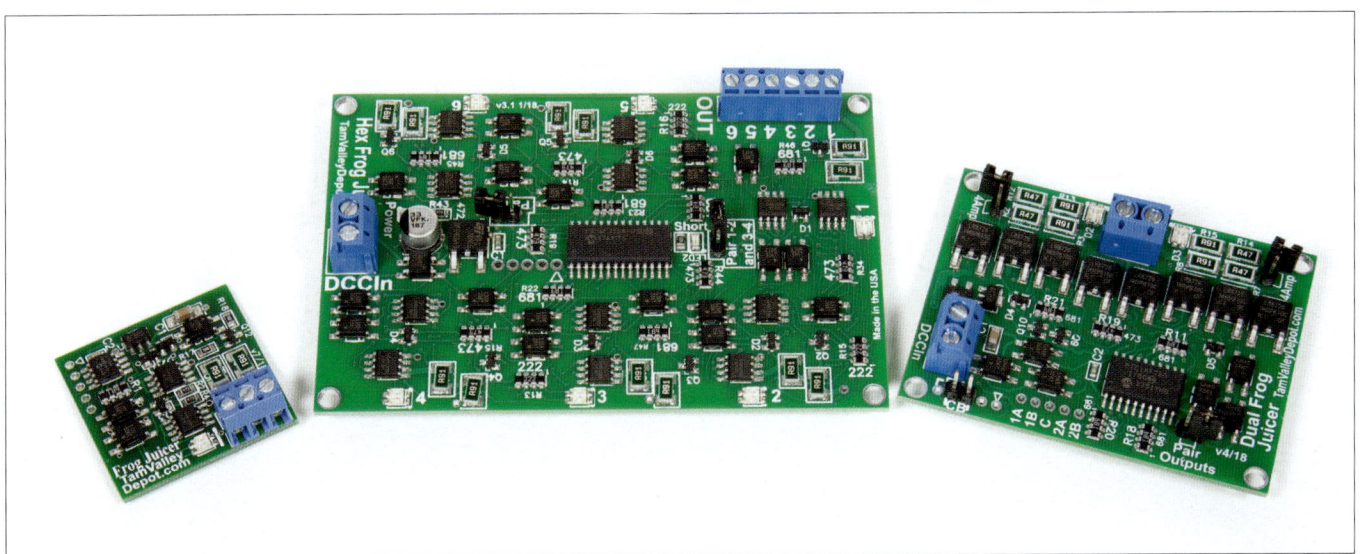

3. Tam Valley's Frog Juicers are electronic circuits that detect short circuits at turnout frogs and automatically switch the polarity before the command station or booster shut down.

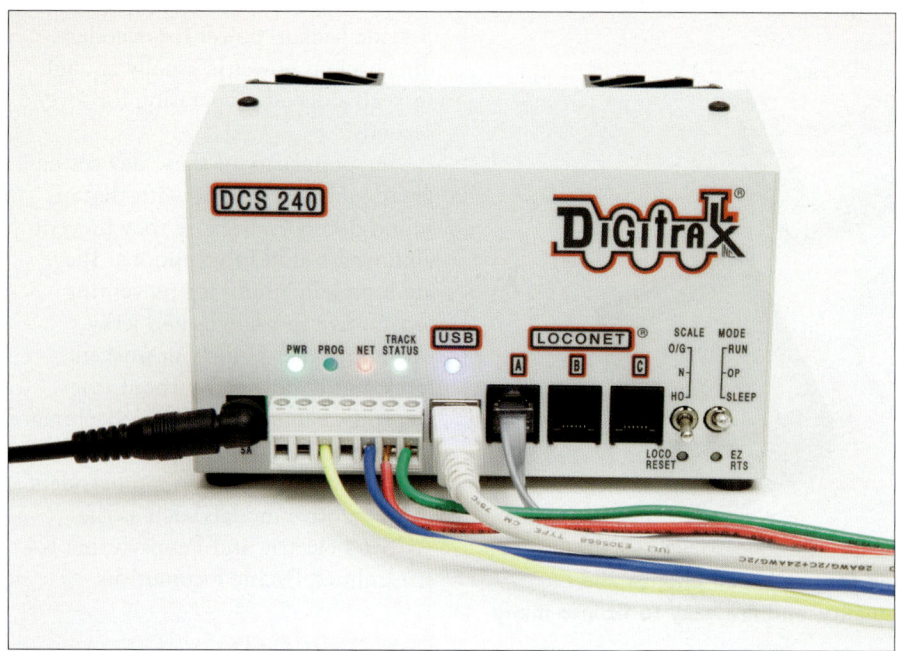

4. Digitrax began revamping its line of command stations in 2016 with the Digitrax DCS240. The system includes the addition of a USB port to allow a computer interface.

before your DCC system shuts down. Frog Juicers are available in mono, dual and hex versions with one, two, and six individual circuits on a board. They offer a quick and easy way to prevent annoying shutdowns due to turnout shorts during operations.

Command stations, throttles, and decoders

As a result of technological advances, manufacturers have been adding new features to their DCC command stations and other equipment. Beginning in the fall of 2016 Digitrax began an overhaul of much of its product line, releasing a command station upgrade with their DCS240 advanced command station, **4**.

One great feature of the DCS240 is the built-in USB computer interface. This greatly simplifies connecting the DCS240 to a computer for use with JMRI DecoderPro or other computer programs. Other important features include the ability to supply either 3, 5, or 8 amps of track power depending on the power supply. This feature is also present in the new DCS210 command station as well as the DB210 and DB220 boosters. The DB220 booster has dual outputs, meaning that with an adequate power supply it can supply up to 8 amps to each of the two track outputs.

As with many of Digitrax's new products, the firmware can be downloaded and installed using your home computer. This capability offers the ability of keeping your DCC command station, boosters, throttles, and other accessories up to date as new features are added.

One of the most interesting new systems, the Digikeijs DRS5000, **5**, appeared in 2018 and offers throttle

5. The Digikeijs DRS5000 was introduced in 2018. It provides a computer-based throttle and accessory interface compatible with several systems (laptop is not included). *Bill Zuback*

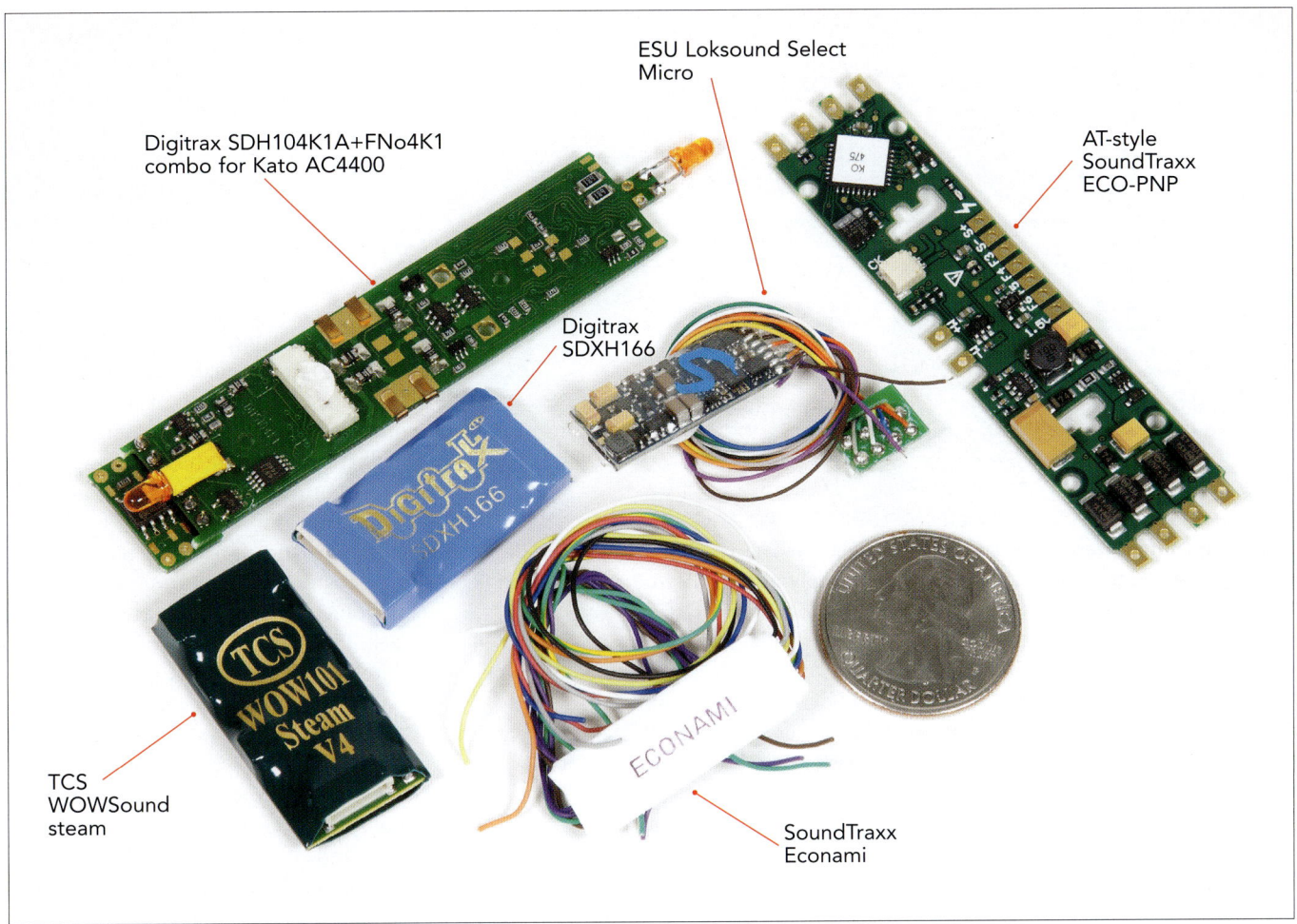

6. Sound decoders continue to increase in features and decrease in size.

and accessory interfaces from Digitrax, Lenz, and any systems compatible with these. For example, EasyDCC wireless receivers can be plugged into the Lenz ExpressNet port to allow use of their wireless throttles. The system also has a built-in WiFi/LAN port and relies on either a Windows 7 or 10 computer system for operation.

Modelers often lament the fact that their throttles are not compatible with other DCC systems, but that is changing. WiFi-based interfaces are now being offered and developed that allow modelers to use different brands of throttles as I just described for the Digikeijs DRS5000 system. Companies like Model Rectifier Corp. and Digitrax (LnWi) offer WiFi interfaces that allow modelers to use their Android and iPhone cell phones to control trains. Cell phone apps like Engine Driver and WiThrottle provide a graphical interface that can control locomotive speed, direction, sounds, and lights. WiFi-capable throttles made by Piko (and soon to be released by TCS) offer the option of a more-conventional handheld device.

Several companies have totally overhauled their sound decoder selections since 2014, **6**. SoundTraxx replaced its Tsunami decoders with the lower-priced Econami decoders along with the upgraded Tsunami2 decoders. The Econami decoders provide a limited selection of sounds and features while the Tsunami2 is available with a wider array of prime movers and features. All are available in an assortment of formats and power ratings.

While the TCS WOWSound steam decoders became available several years ago, the company released a diesel version in 2015 and has greatly expanded the selection of sounds and features on a regular basis since then. The exciting aspect of these decoders is the large number of sounds available on each decoder type. Whistles, bells, horns, and specific diesel prime movers make it a simple matter to find just the right combination to match your prototype.

Just as we were about to wrap up this book, LokSound announced the release of its version 5 decoders. While maintaining similar sizes and formats, LokSound 5 will be available in two different versions—one for the North American and Australian market, and another that will also support Motorola, Selectix, and MFX/M4 digital systems. Among their many features are 10-channel simultaneous playback, 128MB sound memory, 16-bit processor, 3-watt sound output amplifier, and 4 to 32 ohm impedance speaker support. They will be offered with an 11mm x 15mm sugar cube speaker and enclosure.

Switching to DCC

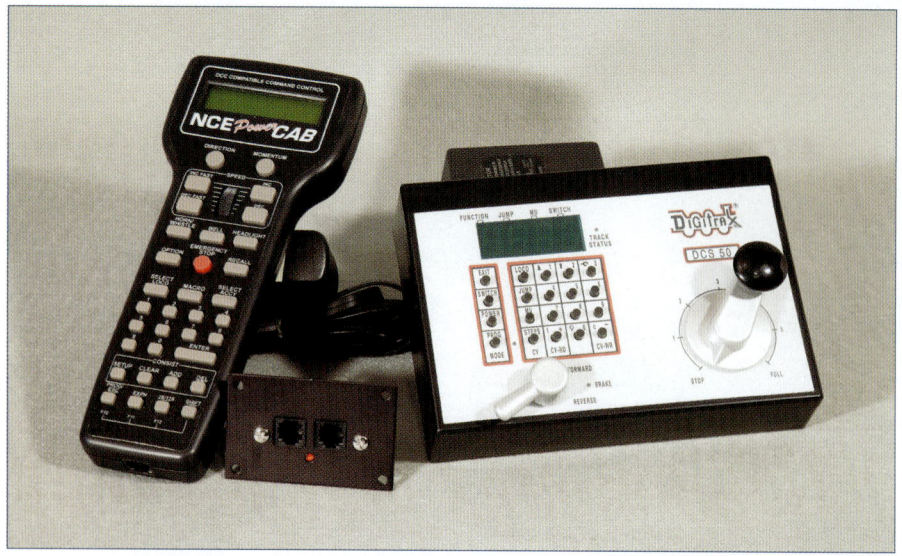

1. Most manufacturers offer introductory systems like the NCE PowerCab (left) and Digitrax Zephyr shown here. They can supply about 1.5 to 3 amps, which is enough power to operate two to six HO scale locomotives.

I got into Digital Command Control (DCC) in 1994 when the first American-made systems began to appear. Over the last 20-plus years, I have seen DCC become one of the most dynamic parts of our hobby, **1**, with several manufacturers offering systems and products. Acceptance of DCC has become so widespread that in many areas it is difficult to find layouts using conventional direct-current (DC) cab control to operate more than a couple trains.

In spite of this, there's still a market for analog locomotives that respond only to changes in track voltage, and the folks at Model Rectifier Corp. continue to make DC power packs, so we know there are still some folks sitting on the fence when it comes to DCC. So let's talk a little about how DCC works, what it offers even the modeler with a small layout, and how easy it can be to convert.

How DCC works

As the name implies, DCC is a type of command control, allowing you to control multiple locomotives on the same piece of track without the need for electrically isolated blocks. The big difference with DCC is the fact that it is digital. Earlier command control systems, like CTC-16, were analog. They operated much like radio remote control, limiting them to controlling speed and direction, with initially only 16 channels.

Digital Command Control, on the other hand, uses a type of alternating current with square waves fluctuating between about +14 to -14 volts (for HO scale). The widths of the square waves that make up the track power can be varied, representing a series of "0's" and "1's" that are the bits and bytes of the digital signal, **2**. The sophistication of this approach has allowed us to go from controlling 16 locomotives to as many as we can provide power for.

All of the commands to control locomotives are created in the DCC command station. This is essentially a small computer that takes the input from the throttle in your hands and translates it into the DCC signal. This command signal is then sent to a booster that increases it to the full voltage that goes out to the track. DCC systems support 4-digit addressing, allowing you to use up to 10,000 different addresses (0-9,999).

But that doesn't mean you can run that many trains—that depends on the capacity of your command station and how many boosters you have. Some command stations can keep track of 250 locomotives, including the status of all their functions and consists.

Many command stations have a built-in booster, especially in systems designed for the introductory market. Most of these introductory systems can supply about 1.5 to 3 amps, which is enough power to operate two to six HO scale locomotives, **1**. If you need more power, additional boosters can be added, and they are available with output up to 10 amps. While some systems offer built-in throttles, most are separate units. Many offer wireless control. Throttles not only serve to control locomotive speed and functions, but some also serve as the programming interface to your command station.

The DCC signal is sent by the command station embedded in track power. It's received by a decoder in the locomotive and translated into instructions. Decoders have powerful microprocessors capable of an astounding number of operations. For example, SoundTraxx Tsunami sound decoders, **3**, have 32-bit

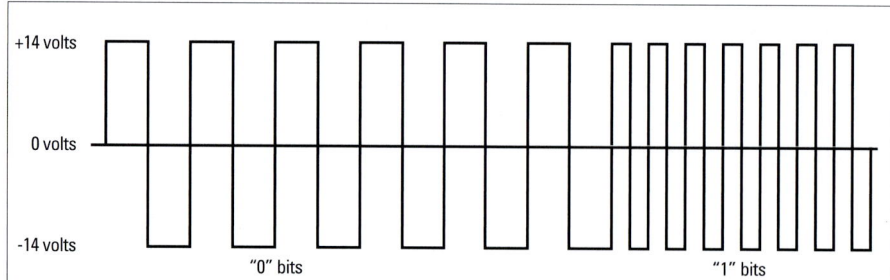

2. The DCC command signal consists of a series of high-frequency square waves that represent the "0's" and "1's" that are the bits and bytes of the digital signal. Having the DCC commands as part of track power instead of just as a small signal riding on top of it increases the signal strength while decreasing interference and noise problems.

microprocessors capable of executing 120 million instructions per second! Decoders have what is essentially a small power pack built into them that takes DCC track power and converts it to DC to run the motor and functions.

Advantages

What else does DCC offer? Multiple locomotives can be combined into consists that respond to a single throttle, making multiple-unit operations smooth and reliable. Helpers can be added to the consist and then cut out easily. You can turn lights on and off, ring bells, blow whistles and horns, and control more functions than most people can keep track of.

We now have support for headlights plus up to 28 functions. The TCS WOWSound decoder in **3** has 18 programmed functions and a total of 30 sounds, some that play at pre-programmed or random intervals. And function controls are not limited to locomotives—you can install decoders in cabooses and passenger cars to control their lights, and SoundTraxx offers its SoundCar decoder for use in rolling stock (see page 78). You can also use accessory decoders to throw turnouts and operate animated accessories on the layout.

If you like to squeeze as much performance out of your locomotives as possible, then DCC has a lot to offer there, too. When DCC first appeared, there were only 14 speed steps available. Manufacturers quickly upped that to 28, and then 128, allowing very precise speed control. You can also set up customized speed tables designed to make your locomotive perform just like the real thing. This and other features make speed matching your locomotives easier.

Operationally, DCC provides a lot of flexibility, especially to the owners of small layouts. This is because a small layout can be wired and operated as one large block using DCC, whereas with DC, you must have at least two blocks for every train you want to run. On a small layout, that can mean fairly short blocks and flipping a lot of toggle switches just to operate a couple

3. Sound decoders keep getting smaller while adding more features. The Train Control Systems WOWSound decoder (center) has a Keep-Alive circuit attached to it to maintain power over dirty track. The decoder on the right is a SoundTraxx steam sound decoder and on the left is SoundTraxx's SoundCar decoder.

trains. With DCC, there are no block switches to worry about.

A lot of folks who are still making up their minds are concerned that converting might be too big a job or too expensive. Assuming your layout is adequately wired, converting to DCC can be as easy as throwing all your cab selector switches to the same cab and hooking up the booster wires to that cab. Yes, there are other things that can be done, but those are not critical and can be done when time and money permit.

The main job is getting your locomotives ready for DCC. Most recent locomotives have a factory-installed socket into which a DCC decoder can be plugged. And the cost of basic, non-sound decoders has come down considerably, with some available for as little as $16. This book includes several installation projects in various kinds of locomotives to get you moving in the right direction. However, if you're just starting to expand your locomotive fleet, then purchasing locomotives with factory-installed decoders is an option.

Finally, there's the cost of the actual DCC system. Many companies offer a starter system that will provide enough capacity to get you going. These introductory systems cost about the same as a single high-end power pack, and for small layouts, they might be all you ever need. They can later be upgraded to a more powerful system as your needs increase. If you have a large layout and run a lot of trains, then you may need a couple boosters, but don't go overboard at first; get a system that can grow with your needs.

Most importantly, DCC operates under standards developed by the National Model Railroad Association (NMRA). These standards specify the electronic characteristics of the DCC signal and the required minimum abilities of decoders to decode them, along with functions, programming, configuration variables, and even wire colors. Decoders that I installed in locomotives 20 years ago still operate with DCC systems made today, and I expect them to still be compatible with systems 20 years from now.

I hope this convinces those of you who have been contemplating DCC to get off the fence and on the road to discovering the increased enjoyment that DCC can bring to your model railroad operations. This book includes many tips and projects to help you toward that end.

1 Selecting the right DCC system

I can't even begin to remember how many times folks have asked me "which DCC system should I buy?" or "which one is best for me?" The answer isn't quite as simple as you might think, as there are a number of things to consider when selecting a DCC system.

One thing you don't have to worry about, though, is compatibility with accessory and mobile decoders made by other manufacturers. Thanks to the National Model Railroad Association's (NMRA) DCC standards and recommended practices, that compatibility is guaranteed. Let's take a look, then, at what does matter.

Choices

First, most systems available today work well, have good track records, and are available at a basic level with a similar array of features. Accordingly, I suggest finding out what the majority of your friends and the local club(s) use. This is important since it means you will have a local support system of experienced users when you need help understanding how some features work.

While most DCC equipment is highly reliable (I have only had to send one throttle back for repair in 24 years), the manuals can leave you scratching your head. This is especially true when you're new to the technology and its jargon. Having someone to translate can relieve a lot of frustration.

Most manufacturers offer an introductory system to help ease you into the technology. Some can

1. Introductory systems like the Bachmann E-Z Command, left, and NCE PowerCab are inexpensive ways to get into Digital Command Control.

be added onto, some may be used as components of larger systems, and some are only suited for solo use. Introductory systems typically are low powered, have fewer features, and allow you to operate fewer locomotives.

For example, the Bachmann E-Z Command DCC system, **1**, comes with a 1 amp power supply that can power two HO or four N scale locomotives (assuming .5 amps per locomotive for HO and .25 amps for N scale). The NCE PowerCab, on the other hand, is rated at 2 amps and therefore can power four HO and eight N scale locomotives.

The Bachmann system supports 10 simultaneous trains while the NCE system supports 12. However, you're still limited by the available amperage, so these are just theoretical values

unless you add more power. Both offer booster units to provide that additional power, which also allows these systems to be used with the larger O and G scales. One advantage of the PowerCab, for anyone needing future expansion capacity, is the ability to use it with NCE's more powerful and advanced Power Pro system.

Many manufacturers also offer more advanced systems, **2**, that allow users to operate more locomotives, control accessories, set up routes, and communicate with computers. With four-digit address capabilities you can enter addresses up to 9999, and large internal memories allow you to store the addresses of hundreds of locomotives and consists. While you may never need that many locomotive addresses, large clubs and big operators might.

However, it also means even the average modeler will not run into memory issues when attempting to use the full system capabilities on a model railroad. These advanced systems typically have more power, with 5 amps being the norm.

Throttles

Throttles are another important consideration. Your throttle is the interface for your DCC system, just like a keyboard allows you to interact with your computer. Because you'll be using the throttle to program decoders,

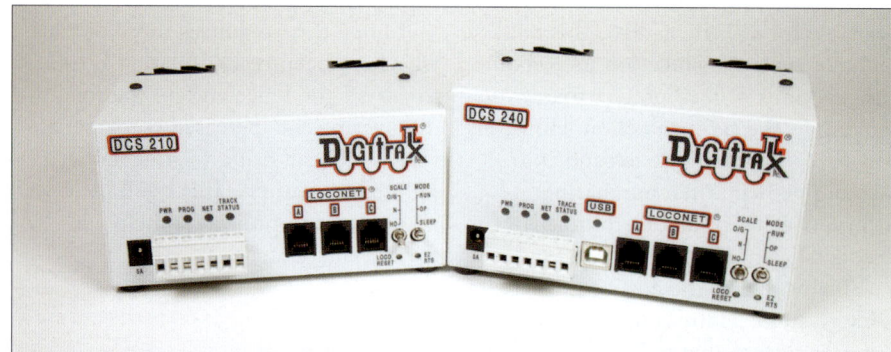

2. Digitrax offers the DCS210, left, as part of an intermediate-level introductory system that probably will fit the needs of most model railroads as far as power and memory, while the DCS240 offers advanced capabilities such as a built-in USB computer interface.

modify system settings, and control locomotives, it needs to be something you're comfortable with. For this reason I recommend visiting a local club or DCC system owner to get a little hands-on time to see how different throttles feel and operate.

There are two basic types of throttles: programming throttles with either vertical knobs or horizontal thumbwheels for speed control, **3**, and small throttles about the size of your palm with a speed control knob, **4**.

The large programming throttles have a variety of push buttons designed for data entry as well as function control. The three throttles in **3**, although similar in the number of push buttons and functionality, each have their own approach to locomotive control.

The Digitrax DT500 has two small speed control knobs, making it simple to control two trains at the same time. The Model Rectifier Corp. (MRC) throttle has a single medium-sized control knob, while the NCE throttle has a horizontal thumbwheel for speed control.

The NCE approach makes it easy to hold the throttle in your hand and use your thumb to control locomotive speed while freeing your other hand for holding a schedule or other operating aids. All of these throttles have an address recall capability allowing you to easily switch from one locomotive to another.

The smaller utility throttles have fewer buttons, since they're not used for programming and other advanced features. Most of the operators I know like the feel of a large diameter control knob, and a unit that fits comfortably in your hand can be an important factor during a three-hour operating session.

Utility throttles are also less expensive than the larger throttles, making them ideal to keep on hand for visitors and operators who don't already own a throttle. Also, because utility throttles don't have the advanced push buttons and capabilities, you can feel relatively safe to just hand off a train to children or drop-in visitors

3. More advanced throttles, from left, like the Model Rectifier Corp. (MRC) Prodigy series, Digitrax DT500, and NCE PowerCab provide full programming capabilities. The MRC has one mid-size locomotive control knob, the Digitrax throttle has two small knobs, and the NCE has a horizontal thumbwheel. In addition, the throttles have individual push buttons to increase and decrease speed.

without fear they'll reset your system or reprogram a locomotive.

Wireless throttles may not be an important item on your checklist at first, but once you've operated on a walk-around layout with one, you're going to want your own. Fortunately, most manufacturers offer some type of wireless throttle capability. These technologies include infrared, radio, and WiFi-based systems. With some systems you can even use an iPhone, iPad, or Android phone or tablet to control your locomotives.

A final consideration is the availability of devices like accessory decoders, block occupancy detectors, computer interfaces, and other accessories. While many of these can be used with any brand of DCC system, some of the extended capabilities are system dependent.

For example, many Digitrax accessories make use of the company's LocoNet communication network to provide feedback to the main unit (command station) and the computer interface. Although you may not plan on using these features now, if at some time in the future you should decide to use some of them, having this

communication capability may be an important consideration.

Selecting a DCC system, especially when you're new to model railroading or are just beginning to build a model railroad, may seem a bit challenging. Taking a little time to plan out what you might want in the future can save you some expense and frustration down the road.

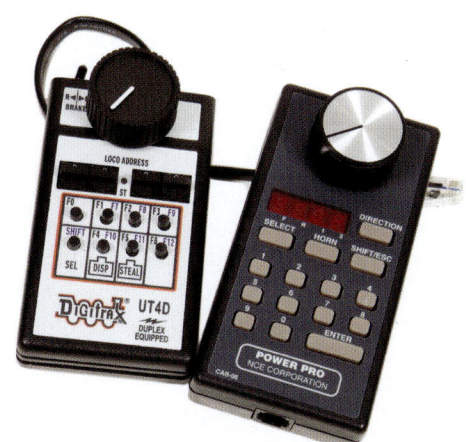

4. Utility throttles like the Digitrax UT4D, left, and NCE Cab-06 have a single large locomotive control knob. These give you the ability to control functions, but can't be used for programming decoders.

2 Layout wiring for DCC

1. A coin large enough to cover both rails makes a good portable short for testing your track. Place the coin on the track, and if the booster shuts down, your wiring is adequate.

One of the big advantages of DCC over conventional direct-current (DC) control is the simplification in wiring. However, this simplification is accompanied by a higher-amperage power feed that requires some different wiring approaches. Although DC power packs typically only put out about 1 amp, many DCC boosters are capable of putting out 5 amps. A basic rule of electricity is that higher amperage requires heavier wires. One reason for this is that amperage and resistance can interact to create heat.

How much heat? Well, to put this in perspective, electrical engineer Mark Gurries recently reminded me that a 15-watt soldering iron can melt solder. (Solder typically melts around 370 degrees Fahrenheit.) A typical 5-amp DCC booster operating at 14 volts can put out 70 watts, or the equivalent of a much bigger soldering iron, so a short circuit can be a potential safety issue.

Although a sustained short could result in a fire, I've only heard of that happening with much larger 10-amp boosters. Shorts are more likely to result in damage to locomotives or rolling stock when equipment derails. Fortunately, DCC boosters have a built-in circuit breaker designed to shut the unit down when a short occurs, and in most cases they operate flawlessly—as long as the wiring is adequate.

Boosters detect short circuits by constantly monitoring the amperage demanded of them by all the locomotives and other loads on a layout. If the total load exceeds the booster's maximum capacity, it shuts down and keeps cycling on and off until the load drops to an acceptable level. However, if there's enough resistance in the wiring and track, the

2. The RRampMeter from DCC Specialties is designed to measure DCC high-frequency track voltage and amperage.

booster won't detect the short and it won't shut down. This is a problem because the booster will continue to supply its full 70 watts to the track until you notice the smell of plastic melting or burning, and shut it off. So just what is adequate wiring and how do we test it to find out?

Testing

First, let's look at how to test your wiring and track. You may have everything you need right in your pocket, something I call a portable short, **1**. Pull out a quarter or larger coin and lay it on the track—the booster should buzz, flash lights, or in some way indicate a short on the track and shut down. Assuming this happens, then your wiring is adequate and you can move your portable short farther along the track. Keep doing this until you've checked out your entire layout. If you find any places where the booster doesn't shut down, then you need to go back and try to find out why.

Another side effect of increased wire and track resistance is reduced voltage on the track. You can see the effects of this when locomotives slow down and speed up by themselves in various sections of a layout. Measuring changes in track voltage around the layout is another good tool for diagnosing track resistance. However, because DCC power frequency is so high (about 8-10 KHz compared to 60 Hz for home AC), standard voltmeters can't reliably measure it.

For this I use a RRampMeter, a device specifically designed by DCC Specialties (www.dccspecialties.com) to measure DCC voltage and amperage. The RRampMeter can be used two ways. By placing the contacts directly on the rails, you can measure track voltage anywhere on the layout, **2**. By moving it around the layout, you can find and isolate areas where voltage drops, and then fine tune your track and wiring resistance. The RRampMeter can also be wired in line with the track bus to provide a constant readout of the voltage and amperage being supplied to the track.

What is adequate wiring? Now that you know how to test your wiring and track, let's talk about what you need to limit resistance. In **3** I've compiled calculations of the voltage drop you can expect per foot of copper wire of various gauges. These are based on 1-wire calculations, so the actual voltage drop will double for a 2-wire track bus. This is due to the fact that each of the two bus wires is acting as a resistor, so the total voltage drop will be equal to the sum of what's lost in each wire on the path from and back to the booster, and the same is true for the rails.

The good news in this table is that voltage drops (and therefore resistance) only become a problem when using small wires or on very long runs at high amperages. Because operating loads generally increase with increasing locomotive scale, you can use smaller wire with N scale than with HO or O scale. I've provided some guidelines for wire gauges by scale in **4**. The largest wire gauges in each scale would be used in longer runs and with larger boosters.

The real problem becomes evident when we look at voltage drops for nickel silver rail, **5**. Again, these values are per foot of single rail, so they double on our layouts. Note that with the typical 5-amp booster and code 83 rail we'd expect to lose up to 0.84 volts per foot, which amounts to 8.4 volts in 10 feet of track! And, as you can see, things get much worse with smaller rail sizes.

Why is rail such a problem? Nickel silver rail is an alloy of about 60 percent copper, 20 percent nickel, and 20 percent zinc. You might think that with that much copper it would be a good conductor of electricity, but it is not: It has a resistivity almost 20 times greater than copper. To compensate for this greater resistance, all you have to do is place your track feeders at closer intervals than you would with code 100 track.

Under normal operating conditions, our loads usually are significantly less than the full amperage output of the booster, so you might think you could

Fig. 3 Voltage drops in wire

Voltage drops per foot for various wire gauges at 1 and 5 amps

Gauge	1 amp	5 amps
12	0.003V	0.016V
14	0.005	0.025
16	0.008	0.040
18	0.013	0.064
20	0.020	0.100
22	0.032	0.160

Fig. 4 Bus wire sizes

Bus wire sizes by scale

Scale	Wire gauge
O	12-10
S, HO	16-12
N, Z	18-16

Fig. 5 Voltage drops in track

Voltage drops per foot per rail for various nickel silver track codes at 1 and 5 amps

Code	1 amp	5 amps
100	0.056V	0.28V
83	0.085	0.42
70	0.152	0.76
55	0.220	1.11

get by with smaller diameter wires. However, because we want to make sure the booster will shut down when a short occurs, we have to plan and wire for the worst case scenarios. So, on my HO layout I use 14 gauge wire for my power buses, with 18 gauge track feeders kept as close to 1 foot in length as possible. I place my feeders at 6-foot intervals on code 83 track and 4-foot intervals for code 70 track.

On really long runs, I often move up to 12 gauge bus wires. Using these standards, I've never had problems with boosters not shutting down. I hope this information will help you with planning or diagnosing the track and wiring on your layout.

2 Power booster basics

In "Selecting a DCC System" (page 10) I touched on system power requirements. Providing power—the role of a booster—is vital, and what you choose will vary based on the size and style of your layout. Now let's look at boosters: what they do, how to determine how many you need, and wiring options. But first you need to understand what a command station does.

The command station actually has two major functions: It acts as a small computer keeping track of your locomotives, function settings, speed, and more. It also converts the commands it needs to send out to the decoders into a digital signal. The digital signal it creates is low voltage and low amperage. The signal needs to be converted to the higher voltage and amperage signal that is sent out on the track; that's the job of the booster.

All DCC command stations I'm aware of have a built-in booster, and that may be all you'll ever need. I regularly operate on a friend's mid-sized layout that runs on the output from a single Digitrax Zephyr unit capable of putting out about 12.8V at 2.5 amps. However, if he wanted to run consists with multiple locomotives, or more trains simultaneously, he would probably need to add a booster.

Variables

When adding boosters to your existing command station, there are a few factors to consider. First, make sure you can easily adjust the output voltages to match. Some command stations and boosters are voltage followers—in other words, their output is close to the input voltage. However, some boosters and command stations have external switches, **1**, for broad adjustments and internal potentiometers that allow fine tuning of this setting. The new Digitrax command stations and boosters allow this adjustment using an advanced throttle.

Boosters are available in various amperages. It's important to realize it isn't necessarily better to use the highest amperage unit available. Instead, several smaller units or power

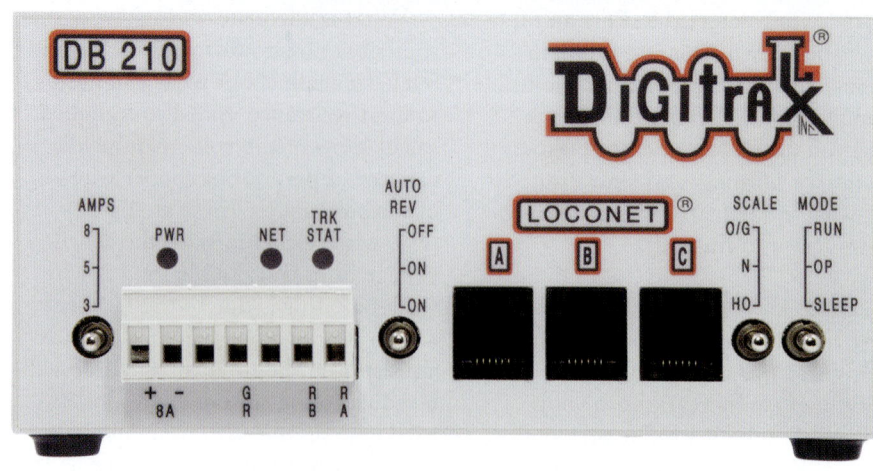

1. The Digitrax DB210 booster can supply 3, 5, or 8 amps depending on the power supply used with it. Note the toggle switch for selecting from three possible voltage ranges for N, HO, and O/G scales.

managers may be a better option. Also, mixing units with different output capabilities isn't a good idea. In some circumstances, such as when a locomotive derails and shorts out across track gaps powered by two different boosters, it's possible for a higher-powered booster to feed current back into the lower-powered unit, possibly causing it to fail. This is especially true if the boosters are from different manufacturers. In most cases both boosters' self-protection circuits would kick in, but it's best to plan for the worst-case scenario.

So how many boosters do you need? That depends on how many locomotives, lighted cars, and accessories you plan to operate using track power. First, consider the locomotives and scale. Motors have changed a lot over the last 30 years, and the amount of current they require has decreased greatly. Consequently, if you have a collection of older locomotives or model in scales larger than HO, motors may require as much as 1.5 amps, each. Modern HO locomotive motors, on the other hand, typically require .3 to .5 amps, and the smaller N and Z scale models draw half that.

Now just because a motor only draws .3 amps, it doesn't necessarily

2. The Model Rectifier Corp. boosters are available in 3.5 and 8 amp versions but externally are essentially the same. Both include a fan to help keep the components running cool.

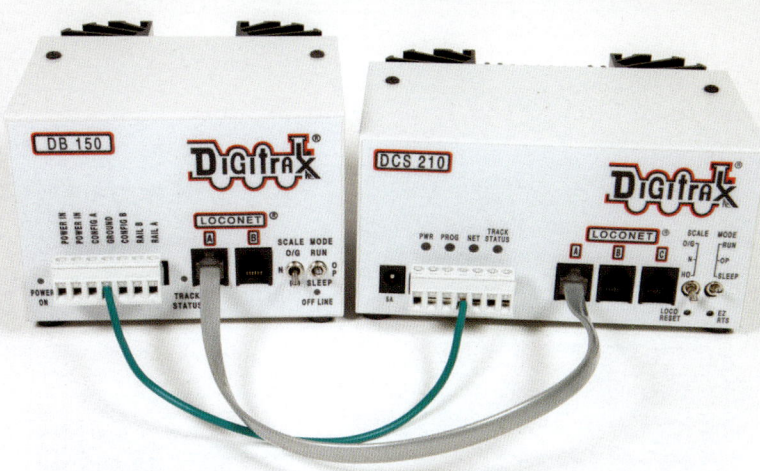

3. Both NCE and Digitrax recommend installing a ground connection (the green wire) between the command station and the booster(s). The flat gray cable allows the DCC signal to be passed from the command station to the booster.

mean that's all you'll need, especially if you're operating diesel locomotives. While steam locomotives typically operated by themselves, diesels are often operated in multi-unit consists. If you want to operate a two- or three-unit consist, you'll need enough power to run all of the locomotives at the same time.

Another factor to consider is if there are any other powered accessories in the train, such as a sound or smoke unit, or in the case of passenger trains, lighted cars. The power demands of those devices have to be added into the total power needs.

As an example, let's look at a five-car passenger train pulled by a pair of diesel locomotives with sound units installed. Assuming these are fairly new locomotives, they probably have efficient can motors requiring less than .5 amp each at maximum load with sound and lights turned on. Next, consider that the five cars are lighted with two .03 amp incandescent bulbs each.

Add that all up and we find that 1.3 amps would be required. Now if you also wanted to hook up a string of lights for the station and stores in town along with some other accessories, you could end up needing 2 amps or more just to operate one train on your layout.

Challenges

Boosters share a common weakness: They generate heat internally. This heat is a by-product of the electrical components used in the circuits and the need to decrease the input voltage to the amount we want out on the tracks. Under normal operating conditions, most boosters can run all day without heat being an issue.

However, if boosters are placed next to a heat source, in a sunny location, or run for long periods near their maximum operating amperage, the excess heat starts to build up inside the enclosure. For this reason some manufactures like MRC install small fans in their boosters to cool them, **2**. Others attach their power transistors to the inside of the enclosure and use it as a heat sink, or install a large heat sink on the enclosure itself, which you can see on the backs of the Digitrax units in **3**.

But in heavy use situations, even boosters with large heat sinks may not be able to dissipate all the excess heat being generated. To prevent damage to their components, boosters have temperature sensors that automatically shut the booster down when the internal temperature exceeds a set value. That's why most 5 amp boosters can only maintain a sustained output of about 3 amps.

If you experience periodic shutdowns when operating several locomotives, then you should suspect excess heat as the cause. The easiest way to deal with this issue is to place a small fan so it blows directly on the booster. If you have all your boosters in one location, you can build a box for them and install a small computer fan in it to keep the boosters cool. You can see my installation on page 19.

Once you add a booster, you'll need wires to carry the command signal between the command station and your booster—the command bus. Most systems use flat telephone-type cables for the command bus. They're available commercially or you can make your own. In many cases you'll also need a wire that serves as a reference ground between the booster(s) and the command station, **3**. Typically, a single heavy copper wire will handle that job. As always, do what the manufacturer recommends.

Also, as soon as you add a booster you'll need to subdivide the tracks into electrically isolated blocks, **4**. Doing this prevents short circuits in one block from shutting down the booster and locomotives operating in the other block. Shorts are a fact of life, especially with some types of turnouts, and boosters come with built-in short-circuit protection so they shut down when a short occurs. That means every train in the block that booster powers will also shut down until the short is cleared.

You can add additional blocks with separate boosters to limit the effects of short circuits, or you can install multiple circuit breakers between the booster and the blocks. Several manufacturers offer stand-alone circuit breakers that make it easy to add additional protection and help manage your power needs on medium to large layouts—turn the page to see how.

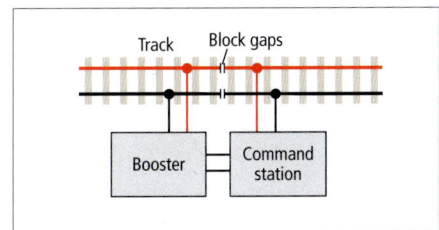

4. When adding boosters to a layout, the track must be subdivided into electrically isolated blocks, each powered by either the command station or booster(s).

Dividing layouts into power districts

How many times have you read or heard that blocks are not required for DCC? For many small layouts that may be true, but for most medium to large layouts, blocks can make operations smoother and more reliable, **1**. What do I mean by a block? Any electrically isolated section of track on a layout can be considered a block. Electrically isolated blocks are created by cutting through both rails at the beginning and end of the desired section of track and providing separate power connections to each block.

But why do we need blocks for DCC? On a direct-current (DC) layout, numerous blocks and related wiring are required to operate more than one train at a time. With DCC you can operate as many trains as your booster can power in one block, so why bother with creating more? The primary reasons for having blocks on a DCC layout are to balance power needs, keep any short circuits from shutting down the whole layout, and to install a signal system.

Yards and industrial switching areas with numerous turnouts are prime targets for short circuits because wheelsets crossing turnout frogs may contact metal track components of opposite polarity. By placing yards and switching areas in their own blocks, you can isolate these potential trouble spots.

The challenge is that as soon as you start creating blocks, you have to power each one. A separate booster for each block is one way to provide power, **2**. But that approach is a bit of overkill and can get expensive.

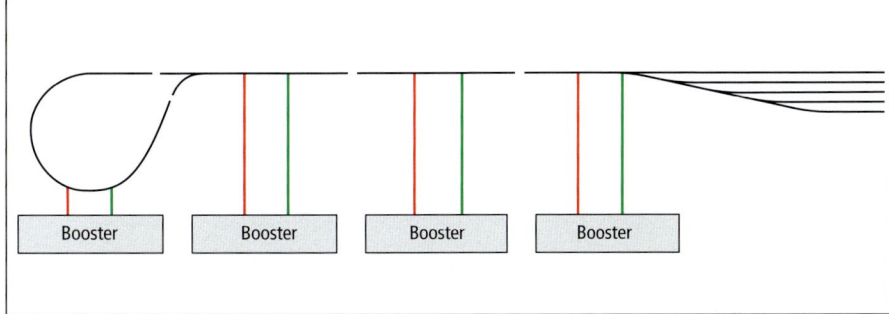

2. One option for powering four isolated blocks on a layout is to power each with a separate booster. This approach works, but it's neither efficient nor ideal and can quickly get expensive.

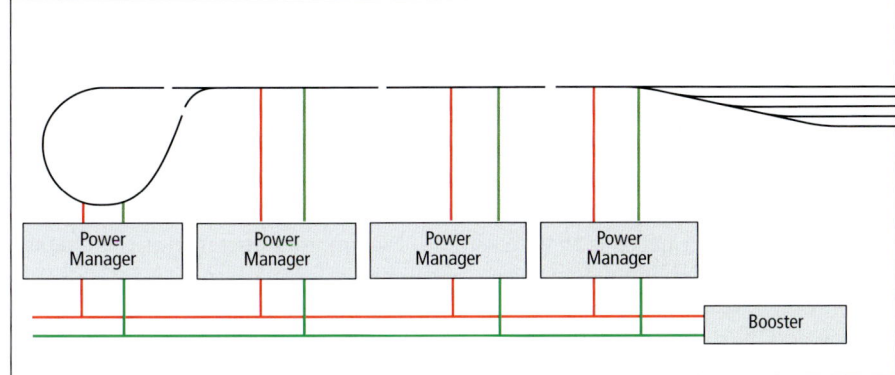

3. A more realistic option for powering four isolated blocks is to use one booster together with four power-management circuits.

1. This isn't exactly what we mean by power management blocks. The goal is to divide model railroads into isolated electrical blocks to limit shutdowns from short circuits and make them run smoothly and reliably.

Power management

Power management allows you to take the power from one booster and divide it among several electrically isolated blocks, **3**. Power-management devices work by assigning a certain amount of current from the booster to each block, with the total actually exceeding that of the booster. The concept is based on the assumption that it's unlikely the maximum amperage will be required in all blocks at the same time.

Let's assume you have four blocks and need a maximum of 2.5A for each one, for a total of 10A. Without power management, you'd either have to buy two 5A boosters, one 10A booster, or a combination of smaller boosters. With power management, you may easily be able to use a single 5- or 8A booster to cover the 10A total.

Balancing power needs

How is this any different from just using one booster to power all the blocks? First, power-management devices provide a circuit breaker for each block. This isolates shorts and also prevents the amperage from exceeding the assigned trip-current value. The circuit breaker in each power-manager circuit will prevent an overload in any one block from shutting down the booster and stopping operations in other blocks.

One limitation to power management is the total amperage being used in all the power-managed blocks at any one time can't exceed the rated capacity of the booster powering them. If you underestimate your power needs or end up running more trains in the power-managed blocks than you expected, you can exceed the booster's maximum rating, and it will shut down.

Developing a well-balanced power-management scheme requires estimating the typical current demand within each block. Keep in mind the age of your locomotives and whether they have sound, a smoke generator, or extra lights. For example, my old Atlas Alco S-2s cruise at about .6A each. But my new Broadway Limited 4-8-4 with lights and sounds on maxes out at about .3A.

Another consideration is whether your passenger cars are lighted or you have accessory decoders drawing power from the track bus. Anything that takes power from the track bus must be added in. One of the best ways to

4. If soldering isn't your specialty, the DCC Specialties PSX-4 is a good option. All you have to do is cut the wires to length, insert them into the connectors, and tighten the screws.

measure your power needs is to wire a DCC Specialties RRampmeter (page 12) into the block bus and measure the average and maximum current draw.

Finally, enter the trip current value into the power manager, and then be prepared to make adjustments in power assignments once you've operated the layout a few times.

Power managers

Several manufacturers offer power-management devices. The two I'm most familiar with and use on my layout are the Digitrax PM42 and the DCC Specialties PSX-4 (**4**). Although the PM42 is designed to integrate with Digitrax equipment, it potentially can be used with any DCC system. The main limitation is that programming must be done with a Digitrax throttle or a computer program that can emulate such a throttle.

Programming is required to change the trip-current value for the blocks (the default is 3A), and to change the speed at which the circuit trips (there are four options), as well as various other functions. The PM42 can be configured for protecting four blocks as either a circuit breaker, automatic polarity reverser, or a combination of the two (although this reduces the total number of blocks that can be managed).

Maximum amperage for the blocks can be set from 1.5 to 12A in 1.5A increments. Although wiring the four outputs may look difficult, I found it fairly straightforward. Good soldering skills and tools make the task more manageable (see page 40).

The PM42 circuit board is powered by a separate 14 to 16V transformer. Should you find that you need more power than is provided by a single booster, you can add another and rebalance the four outputs.

The main limitation of the PM42 is that the same trip-current value applies to all four blocks. You can't have one block set to 3A and the rest set to 1.5A each.

DCC Specialties produces the Power Shield series of power managers, typically referred to as PSX, **4**. The trip current can be set from 1.27 to 17.8A in 1.27A increments and can be programmed using either jumpers soldered to the board or DCC ops mode programming.

Because the PSX circuits are electrically independent, you can set different trip current values for each block. The PSX can be programmed by all the major DCC systems and can interact with their control buses. In addition, instead of having selectable trip speeds, PSX uses an intelligent logic algorithm that can discern the difference between a short and the current load caused by the in-rush current of sound decoders starting up.

One handy feature of the PSX is that it's available in several configurations. This is important when it comes time to decide whether you want your power managers all in one

Boosters of different ratings

The sidebar at right shows a command station and booster in an enclosure. The photo shows I had mixed an 8A DCS200 and a 5A DB150. A question is: Can you mix boosters of different amperages like that? To get the full technical answer on this I posed the question to the electrical engineers at NCE and Digitrax.

Basically, the answer is you're probably better off not mixing boosters of different amperages or from different manufacturers. There are scenarios during short circuits across block gaps where the higher amperage unit can feed back into the lower amperage unit and damage it. However, it's possible to get away with it for a long time under the right circumstances.

First, the boosters would need to be set so they are all putting out essentially the same voltage. Second, the modeler must have done the wiring properly so that both units' short-circuit protection shuts down as designed, and then there may be no problems.

Also, the use of ancillary circuit breakers like the NCE EB1, Digitrax PM42, or DCC Specialties PSX series may help prevent problems should shorts occur. In most cases, if a sustained short occurs across power district gaps due to a derailment, the wiring in the locomotive will likely burn out before any damage to the boosters occurs. This is why you may hear folks say they've been doing this for years without any problem. However, remember the DCC version of Murphy's Law—if it *can* happen, it eventually *will* happen.

Preventing thermal shutdowns

One consequence of using power management is that your boosters may be forced to work harder. This could lead to overheating, since most boosters have two amperage ratings: the advertised maximum rating and the unadvertised sustained operating rating.

A common 5A booster can typically maintain that level of output for fairly short periods, after which it will develop a thermal overload and shut down. Most can only maintain a sustained output of about 3A. This is because under a sustained load the electronics in the enclosed case build up excess heat and exceed the preset overload temperature, usually 122 degrees F.

Heat buildup is dependent on both the load as well as location. If the booster is in a small non-ventilated area or near a heat source, it will shut down faster than in a cool, well-ventilated location. Because the metal case usually acts as a heat sink, units should not be stacked on one another, a transformer, or any other heat source.

With a properly balanced power-management scheme, booster(s) may need to operate closer to their maximum rating more of the time, instead of the lower sustained rating. By having a small fan blow directly on your boosters, you can reduce thermal overloads. However, if you have a couple of boosters, power managers, and other electronic devices, it can be more efficient to put them in an enclosure with a fan to provide additional cooling. This also helps organize your components and wiring, and prevents dust from accumulating on the components.

For my boosters I built an enclosure out of wood and Plexiglas. This is a simple box sized to provide an 8" tall, 11" deep, and 24" long interior. The materials consisted of a 6-foot 1 x 12 cut into four pieces for the top, bottom, and sides and a ¼" x 8¾" x 24¾" sheet of plywood for the back. I used a 9" x 25" piece of Plexiglas for the door and a piano hinge to allow it to swing down. Finally, I added some magnetic latches to hold the door shut.

Airflow is provided by a muffin fan similar to those installed in the back of desktop computers. I chose one that operates on DC power and uses a plug-in transformer for power. The air escapes through the various holes required for all the power and bus wires.

You have the choice of installing the fan so that it blows air in or out. I installed the fan on the inside of the enclosure so it pulls air in. I placed a small circle of furnace filter material in the opening to keep most of the dust out.

The case on many boosters serves as a heat sink. A large, finned heat sink comprises the rear of the enclosure on Digitrax boosters.

Larry built an enclosure for his boosters from Plexiglas and wood. The basic box measures 8" tall, 11" deep, and 24" long. A brass piano hinge makes the Pleixglas door easy to open if any maintenance is required. Magnetic latches securely hold the Plexiglas door shut, which helps keep dust out.

Muffin fans like this one are commonly used to cool power supplies in desktop computers.

Larry installed the muffin fan toward the rear of the enclosure so it blows air on the large, finned heat sinks.

Larry drilled several 1" holes around the box for wiring runs and to allow air to circulate.

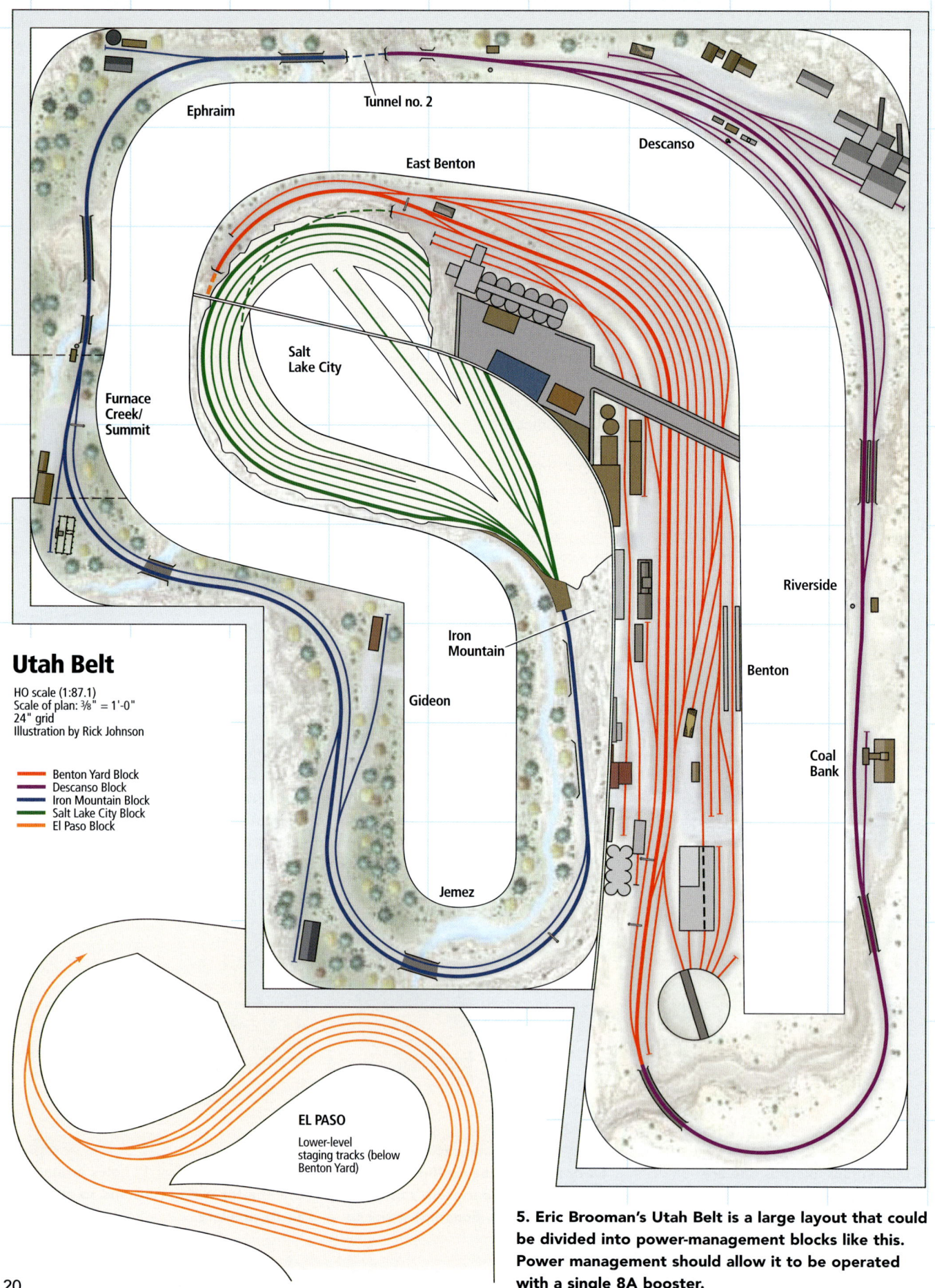

5. Eric Brooman's Utah Belt is a large layout that could be divided into power-management blocks like this. Power management should allow it to be operated with a single 8A booster.

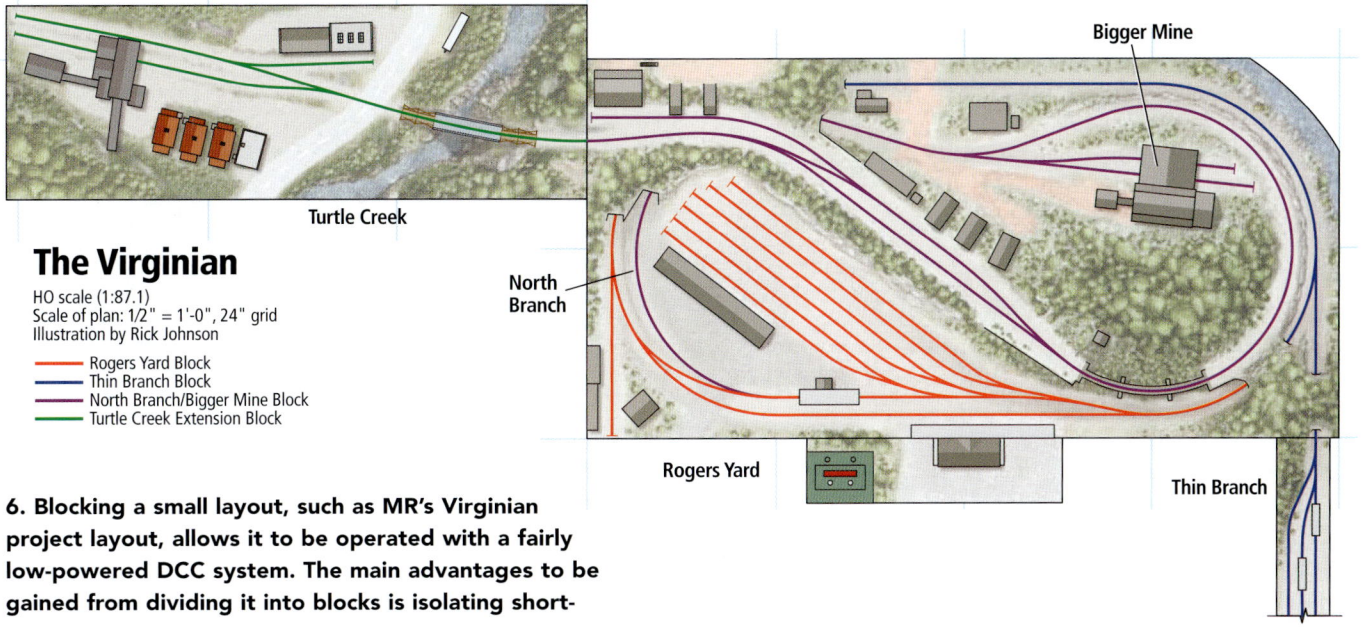

The Virginian
HO scale (1:87.1)
Scale of plan: 1/2" = 1'-0", 24" grid
Illustration by Rick Johnson

— Rogers Yard Block
— Thin Branch Block
— North Branch/Bigger Mine Block
— Turtle Creek Extension Block

6. Blocking a small layout, such as MR's Virginian project layout, allows it to be operated with a fairly low-powered DCC system. The main advantages to be gained from dividing it into blocks is isolating short-prone areas.

place or distributed around the layout near the blocks they power.

The Digitrax PM42 has all four power-manager circuits on a single board, so all your wiring has to fan out from one location. The PSX power managers are available with one, two, three, and four circuits on a single board. You can scatter combinations of them around the layout with their boosters and shorten wiring runs.

Also, because the circuits are completely independent, if you purchase a board with multiple circuits, they can later be separated along the provided score lines if you decide to go from a central to distributed configuration.

There are a couple things to look out for with power managers. First, low-output boosters can create problems, since they may not be powerful enough to stay on if the power manager has to cycle on and off repeatedly during a full short. To counter this problem, the PSX can be configured to remain off in case of a such a short, and then can be restarted using a remote switch. The unit also offers a power boost mode, which helps some boosters start even with heavy sound decoder loads.

Locating blocks

When dividing a model railroad into blocks, I usually start by isolating yards and switching areas that have lots of turnouts. Next, I focus on staging yards. Finally, I block out long runs of single or double track.

As an example, see how power management can be used to full advantage on medium- to large-size layouts like Eric Brooman's original Utah Belt, **5**. I would divide this 325-square-foot layout into five blocks: the main yard at Benton; Descanso and the long section of mainline track associated with it; the long stretch of track from Tunnel no. 2 to Iron Mountain; the hidden reverse loop and staging tracks in Salt Lake City; and the hidden reverse loop and staging tracks in El Paso.

Using a PSX-3, I would start with a current draw of 3.81A for the Benton block and 2.54A each for the Descanso and Tunnel no. 2 blocks.

The two reversing loops require special attention. DCC Specialties also makes the PSX-AR, which adds auto reversing to the PSX circuit board. I'd use two of these set to 1.27A each to control the reverse loops and staging tracks at Salt Lake City and El Paso. Since the total potential current demand for all five blocks is 11.43A, I would start with an 8A booster to power the blocks.

On the 50-square-foot Virginian project layout, which first appeared in the January through June 2012 issues of *Model Railroader*, Model Railroader Video Plus producer David Popp wired it as one large block. He used a 2A NCE PowerCab DCC system capable of simultaneously operating two or three HO locomotives.

Because of the low amperage output of this layout, it's hard to get much advantage from power management. I would use a slightly more powerful 2.5 to 3A system, such as the Digitrax Zephyr, **6**. The best David could have done is divide the layout into four blocks: Rogers Yard, the North Branch and Bigger Mine, the Thin Branch, and the Turtle Creek extension.

Using a PM42, David could have assigned a 1.5A trip current to all four blocks for a total of 6A powered by the 2A booster. This would have prevented a short circuit in one block from shutting down operations on the whole layout.

Power management can be a useful tool for operating your model railroad. It can reduce the frustration resulting from short circuit shutdowns and power overloads. All it takes is a little time to balance the power-manager settings for your specific operating scheme.

2 DCC power supplies

Providing proper power for DCC command stations and boosters has changed dramatically over the past 25 years, and we've experienced three distinct generations of power supplies. In the first generation, during the early days of DCC in the mid-1990s, modelers were pretty much on their own to come up with a power supply that could provide the approximately 15 volts and 5 amps that most DCC systems required. With a couple exceptions, DC power packs just weren't that powerful, and transformers that large were few and far between—I got by with an 18VAC, 4A RadioShack unit.

Then a few companies like Loy's Toys and Springhaven Shops stepped in and began to offer kits with transformers powerful enough to meet the full demand of the DCC systems. Typically, these do-it-yourself (DIY) kits came with a transformer, cord, switch, and fuse holder, **1**, and supplied unregulated AC or DC power. The big missing component was a proper enclosure.

As a result it was common to see all these components mounted on a piece of plywood sitting on a shelf under the layout, with exposed 120 VAC connections right at the same level as curious children. I have seen such arrangements at train shows, clubs, and home layouts.

I still see this kind of setup on occasion today, and it's a major hazard and legal liability. So my first

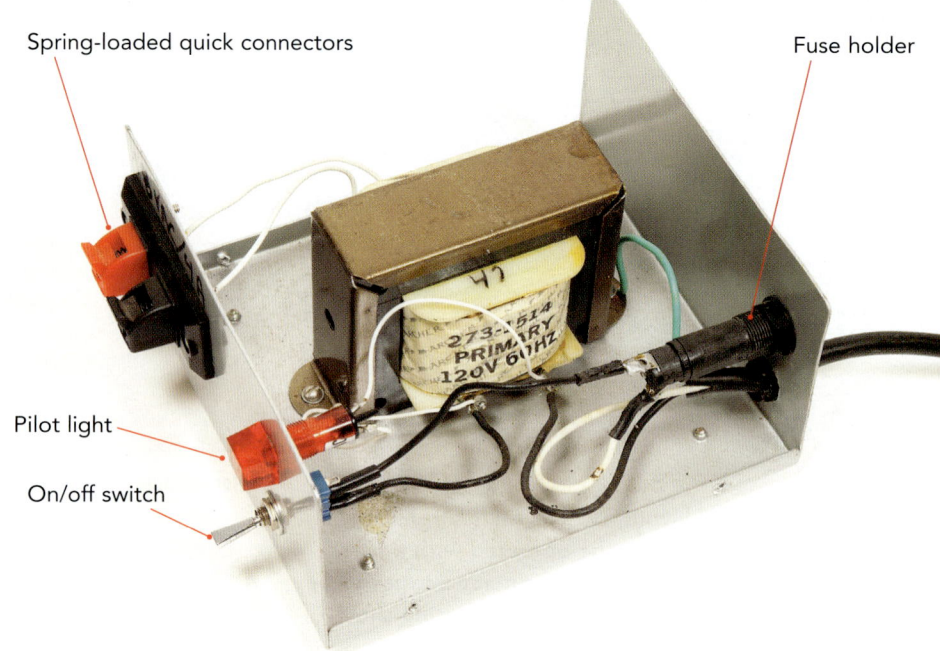

1. My first-generation DCC power supply was a do-it-yourself unit cobbled together with an 18 VAC, 4A transformer from RadioShack. I installed spring-loaded quick connectors, an on/off switch, pilot light, and fuse holder, all in a RadioShack enclosure.

recommendation is if you have such a setup, either buy a safe enclosure, build one as soon as possible, or replace the whole thing with a more-modern power supply.

DIY kits pretty much disappeared from the marketplace when a second generation of power supplies was introduced by DCC manufacturers such as Bachmann, Digitrax, Model Rectifier Corp., and NCE. These typically had either a fuse or circuit breaker and an enclosure, **2**, and were tailored for the manufacturer's systems. Safety was now an issue, so these were Underwriters Laboratories (UL) listed units.

We've begun to see the introduction of the third generation of power supplies. These typically are switching power supplies that produce clean, well-filtered DC power, with a self-resetting fuse or circuit breaker, all in a sealed plastic enclosure. Some can even be set for different voltages, **3**. Digitrax recently overhauled its entire line of boosters and command stations, which now require a clean DC power supply.

So what brought on this third generation? The important thing about switching power supplies compared to the old ones is they're much more efficient. Switching power supplies use transistors to rapidly turn power on and off to alter voltage and current.

Many countries now have energy efficiency standards, and to meet them, manufacturers have been turning to switching power supplies. The power supplies also simplify command stations and boosters because these components no longer need to convert AC to DC at the proper current and voltage.

Most switching power supplies have

2. Model Rectifier Corp. included this second-generation power supply with its original Prodigy Advance DCC system.

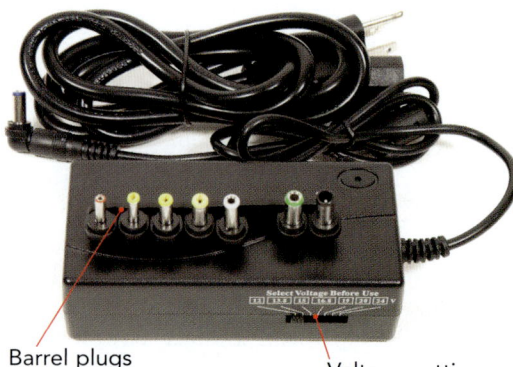

3. Digitrax offers this third-generation switching power supply with seven voltage settings and a selection of eight barrel plugs.

a separate cord that allows them to be used in countries with different plug configurations, plus they usually have a switch or internal circuitry allowing them to be used with 110 to 240 VAC line power.

I've recently seen some modelers use switching power supplies from sources other than DCC manufacturers. Most of these power supplies are made in China and come in perforated steel enclosures, **4**. The downside is they usually don't come with a cord. However, they generally are protected against short circuits, overloads, and over-voltage conditions.

Companies like Digi-Key, Jameco, and Mouser all offer switching power supplies in a variety of voltage and amperage ratings. One thing to be aware of is the different ratings covering them. Some may carry both UL and European Union certification (Conformité Européene, or CE), some may only have CE certification, and some may have no certification at all. This information is usually provided even on eBay listings.

Typical setup

Let's take a look at a typical modern setup. Notice in **5** that all the connections are made using screw terminals along one end of the enclosure. The first three terminals on the left are for the line power and ground wires in the power cord. For these I have a large collection of computer cords left over from my old office—dumpster diving may yet prove profitable!

On my power supply, the black wire is connected to the "L" terminal, the white wire to the "N", and the green wire is for the ground. Next there are two "COM" terminals for the negative DC wire and two "+V" terminals for the positive DC wire. Some units only have one set of DC terminals. Finally, the small yellow potentiometer is for adjusting the DC voltage.

This particular unit came with an amber plastic cover that fits over the screw terminals. However, since it doesn't completely cover the AC line connections, I keep the power supply in an enclosure with the command station/booster and a cooling fan.

Since I'm a bit of a worry wart, and in spite of the fact that the unit has internal overload protections, I still place a 3 amp fuse on the black AC input lead and an 8 amp fuse on the negative DC lead. They may not be necessary, but as Ben Franklin liked to say, "an ounce of prevention is worth a pound of cure."

While we're talking about providing power, let's touch on a facet that's seldom mentioned. Our layouts are usually in basements or spare rooms that are likely served by only one electrical circuit, and they may even share that one circuit with other rooms in the house.

Also, in some houses, the ceiling lights may be on the same circuit as the wall outlets. In the United States, most general-use circuits have a breaker with a 15 amp rating, and the general rule is that no circuit breaker should be loaded to more than 80 percent of the maximum for more than three hours. That means the continuous rating for most circuits should only be about 12 amps.

When you start adding power supplies for all the boosters, lights, and other accessories on your layout, take a little time to add up the amperage load you're installing. However, be aware that an 8 amp power supply doesn't pull that much current from the main.

For example, my 8 amp switching power supply only pulls 2.5 amps, and that's only when it's maxed out. The same goes for your other power supplies. Most now have both the 120 VAC input rating and the DC output rating stamped on the enclosure. So take a few minutes to check your equipment and stay under the safe load rating of your house circuits.

And don't forget the mini fridge stocked with soft drinks, the coffee pot, the soldering iron, maybe a window air conditioner and a dehumidifier, and so on. You'll be amazed how fast all these things add up.

4. This Chinese-made unit is one of my current switching power supplies. It can be powered with either 110 or 220 VAC and comes in a perforated steel enclosure, providing good ventilation.

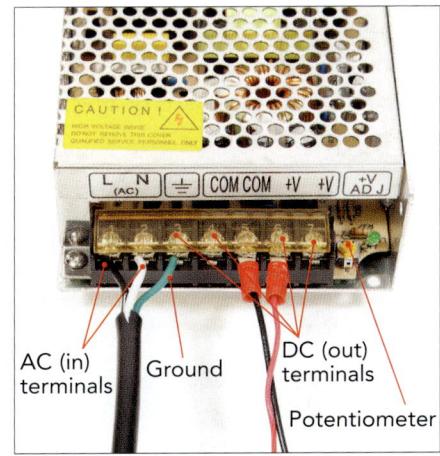

5. The screw terminals placed along one end of the enclosure make connecting the AC and DC wires easy, and the small yellow potentiometer allows you to fine-tune the DC voltage. Note the amber plastic cover over the screw terminals.

2 Improving power and signal reliability

For generations, model railroaders have been working on their locomotives to get them to run smoother—especially at slower speeds—with more power and no stalling. When I first got serious about model railroading back in the 1980s, it was common practice to replace factory-installed motors, flywheels, and universal joints and tinker with mechanisms and gears to improve operation.

Fortunately, most locomotive models today are much improved, with more-efficient motors and smooth drive trains, so we don't generally have to resort to these extreme measures. That said, we still can face undependable operation. So what hasn't changed?

If there's one thing DCC is sensitive to, it's poor electrical pickup. Keep in mind that DCC decoders are essentially little computers, and we all know what happens when power to a computer is interrupted even briefly—it shuts down. The same thing happens to a decoder when it briefly loses power.

And the really bad thing is that when power returns, instead of continuing at its previous speed, the decoder first resets to speed step "0" then jumps back to the previous speed once it receives an update from the command station. This results in jerky operation that in extreme cases can knock cars off the tracks, and also leads to interruption in sound effects with sound decoders.

So, in our search for more reliable operation, we have to turn to the causes of power interruptions. These include unpowered turnout frogs, dirty track and wheels, and unreliable internal connections. Let's take a look at each of these and discuss some easy ways to fix the problems.

Keys to smooth operation

Powered frogs. On pages 30 and 32 I'll discuss turnouts and how to power frogs, so I won't go into detail on that here. If you decide to use electrically dead frogs, there are devices you can add later to deal with stalling. Plus, as long as you use turnouts with metal frogs, you can add feeders later, although it will be a lot more difficult and, in some cases, ugly.

Clean track. Dirty track and wheels are common problems for locomotives with DCC. Dirt and grime on rails comes from airborne dust, scenery materials, oxidation of the rails, and oil and grease from our locomotives. These all mix to form the dark greasy grunge that can cake wheels and rails. The easiest way to prevent it is to keep things clean in the first place.

A lot of dirt and dust filters down from unfinished ceilings and works its way up from raw concrete floors. The best way to limit that is by finishing the train room. It's helpful at minimum to at least paint the floor and add a ceiling (or even staple plastic sheeting to the ceiling joists).

Garages in particular can be dirty places for layouts, and the only real option there is to hang plastic sheeting over the layout and add removable plastic curtains to protect it when not in use, **1**.

Scenery materials can cause problems. It's a good idea to cover track (blue painter's tape works well) when applying plaster, ground foam,

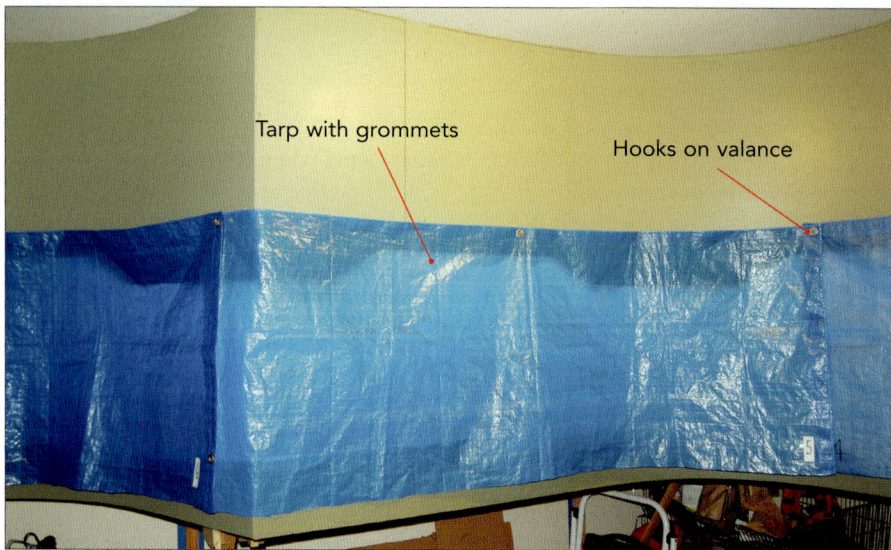

1. In the June 2012 issue of *Model Railroader*, Howard Lloyd described how he made this curtain to protect his layout from dust using a plastic tarp available at most hardware stores. *Howard Lloyd*

2. The metal strip on top of older Athearn motors conducts current from the truck gear towers to the motor, but can be a point of poor electrical contact if the strip sags or the contacts corrode.

static grass, or other scenery materials.

When converting an existing layout to DCC, I strongly recommend a thorough track cleaning. This may require using something like Goo Gone or isopropyl alcohol to remove any built-up deposits on the track. If you do use Goo Gone, it can leave a film behind that will attract more dirt, so always follow up with a clean, dry rag. Isopropyl (rubbing) alcohol, on the other hand, will evaporate off the rails. Just be sure to open the windows when you use a lot of it. Don't use denatured alcohol or lacquer thinner.

Once the rails are clean, a moist rag or track cleaning car will usually be enough to remove any dust that accumulates. When you get around to a thorough cleaning job, don't forget the wheels on all your locomotives and rolling stock. If you don't get it off the wheels, that grunge will just end up back on your clean rails, and you'll have to start cleaning all over again.

Locomotive wiring. Another source of erratic performance can be traced to unreliable electrical connections inside the locomotive itself. Older Athearn "blue box" models and others of similar design are especially susceptible. If you look at the chassis of older models of this type, you'll see a metal strip on top of the motor, **2**. This strip picks up current from the trucks and passes it to the motor.

However, electrical continuity depends on physical contact between the strip and the metal riser on each gear tower, which simply slide over each other. Over time, these strips can rust or sag and electrical continuity may become unreliable. The best fix is to replace the metal strip with flexible copper wire soldered to the gear towers and motor contact.

Another issue with the old design is the contact between the truck and frame, **3**, which conducts power to the motor. The trucks ride on a bolster with a small cast-in pin on the frame that mates with a hole in the truck. For DCC use, the motor must be isolated from the frame.

The bearing surface on the truck consists of a steel plate that can rust

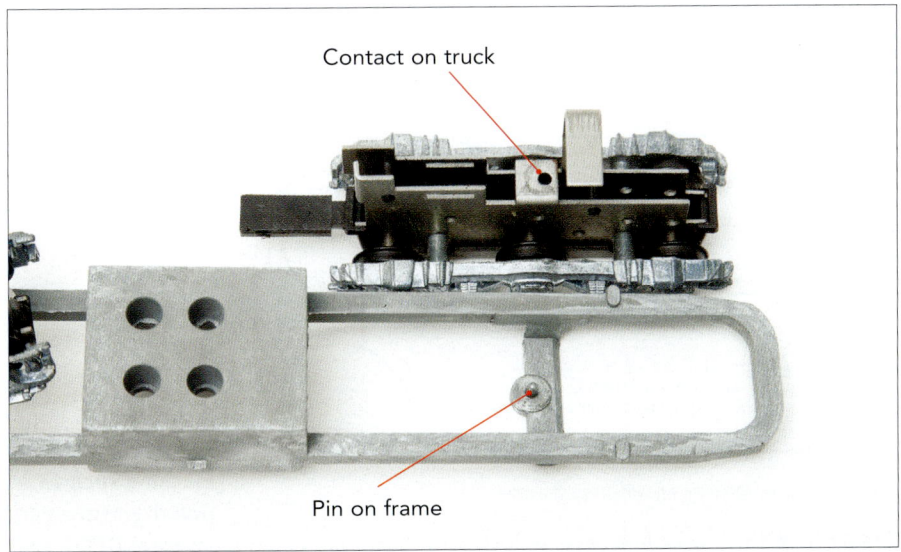

3. The Athearn bolster pin on the frame mates with a hole in the truck's gear case and serves as the other electrical contact point.

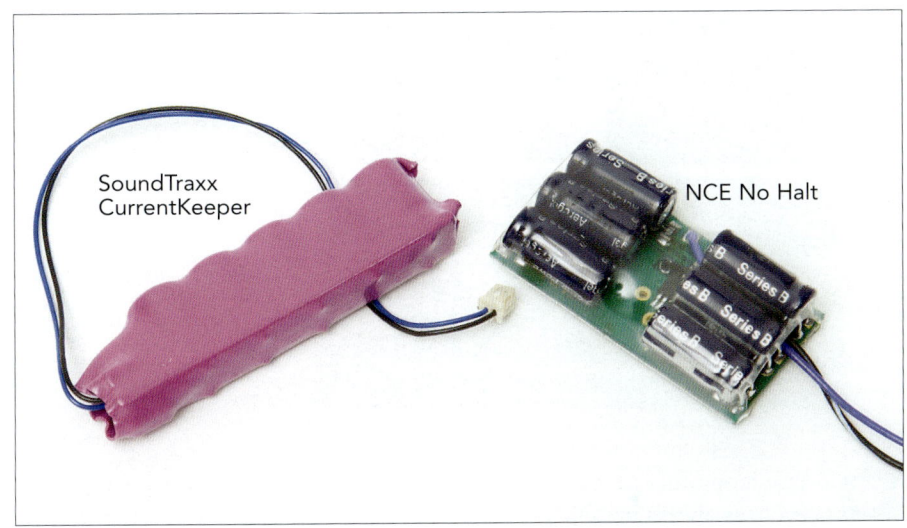

4. Capacitor modules, like these from SoundTraxx and NCE, can power a decoder for several seconds, allowing it to pass uninterrupted over dirty or unpowered sections of track.

or get fouled with lubricants and dirt. This can result in intermittent loss of electrical continuity. The fix here is to solder a piece of flexible copper wire to the truck frames and from there to the motor. All of these fixes can be easily accomplished at the same time as a decoder installation.

Electronic solutions. To prevent stalling, a popular option is to install capacitor units. Most of the major decoder manufacturers offer capacitor units under brand names such as Keep Alive, CurrentKeeper, or Power Xtender for use with their decoders, **4**. These consist of several capacitors that charge themselves while receiving track power and feed that to the decoder, acting essentially as little batteries. When your locomotive hits a dirty section of track or a dead frog, the engine will glide right over it, powered by the capacitors.

Most capacitor units carry a large enough charge to keep a decoder operating for several seconds after track power is turned off. The net result is smooth, uninterrupted performance, including sound. See page 62 for a more-detailed look at these devices.

2 Reverse loops, turntables, and wyes

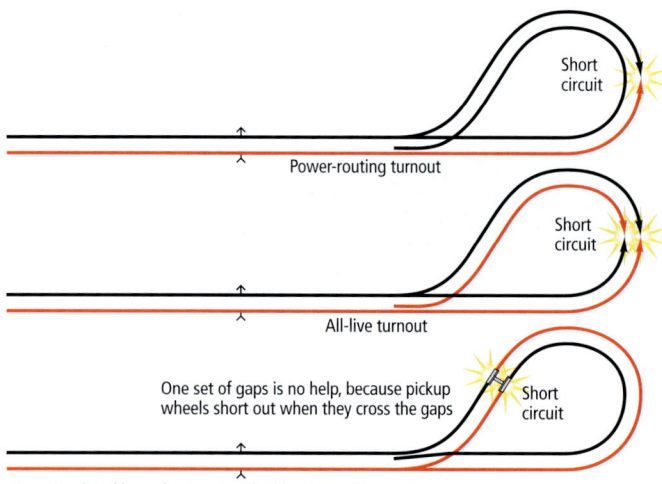

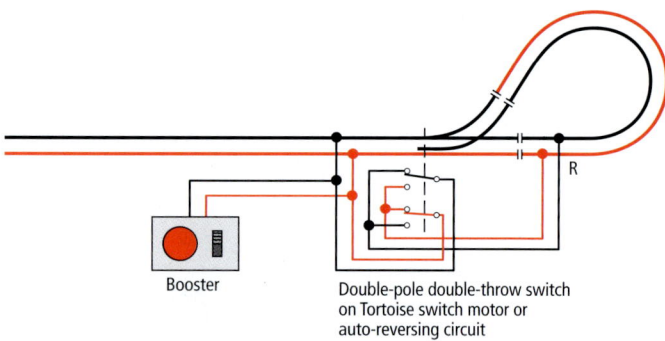

1. Any time the tracks in a reverse loop turn back on themselves, it creates a short circuit.

2. An old DC approach that also works with DCC is to use the double-pole double-throw (DPDT) contacts on a switch machine such as a Tortoise to automatically change the track polarity. However, if you don't use switch machines with integral DPDT switches, an auto-reversing circuit will correct the polarity.

3. Auto-reversing circuits such as the DCC Specialties PSX-AR (bottom), Digitrax AR1 (top left), and Tam Valley Depot Dual Frog Juicer (top right) make correcting track polarity problems easy.

Reverse loops, turntables, and wyes have always presented model railroaders with the challenge of how to maintain correct polarity when turning or reversing a locomotive or a whole train. All of the methods for preventing short circuits in the direct-current (DC) environment should also work with DCC. And the good news is that in most cases there are easier ways to correct the problem on DCC-powered layouts than with DC systems.

Let's take a look at each potential problem, then outline the ways to correct it on a DCC layout.

Reverse loops

The simple reverse loop is probably the most common of these three potential issues. As you can see in **1**, this problem arises any time tracks loop back on themselves, creating a short. The first step in correcting this problem is to isolate a section of track in the reverse loop by cutting gaps in the rails at each end. Note that it isn't enough to simply cut one set of gaps; you have to totally isolate this section. This section of track will serve as the reversing track where the polarity correction takes place. But how do we go about correcting the polarity?

The old DC approach that also works with DCC is to use the double-pole double-throw (DPDT) contacts on a switch motor such as a Circuitron Tortoise to automatically change the track polarity as shown in **2**. Using this method, the reversing track polarity is changed when the turnout is set for the train to exit the reverse loop.

Unlike DC operations, because the locomotive direction of travel is controlled internally by the DCC decoder, the polarity of the reversing track can be changed while the locomotive is in motion, which simplifies operations.

Another option is to replace the DPDT switch with one of the electronic auto-reversing circuits available from DCC Specialties, Digitrax, Model Rectifier Corp., and Tam Valley Depot, **3**. These devices automatically detect when a short circuit occurs and immediately reverse the polarity of the wires to the reversing track. These can be especially useful in areas of complex track or in long reverse loop situations.

One thing to keep in mind is the reversing section of track needs to be at least as long as the longest consist of locomotives that will be run through it. If the section is too short, once the

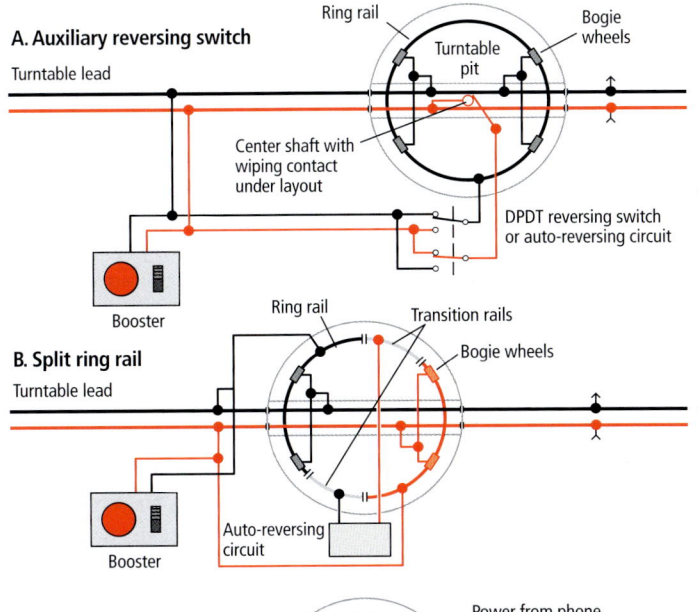

4. Turntable short circuits can be avoided using DPDT toggle switches in the feeders to the turntable tracks. Three common turntable wiring schemes are shown. In all cases the DPDT switch can be replaced with an auto-reversing circuit.

5. As with reverse loops, wyes create short circuits as a result of track and turnout configurations.

6. Correcting the short circuit in a wye can be done with DPDT switches or an auto-reversing circuit.

polarity changes, the wheels on trailing locomotives will cause short circuits when they bridge the gaps. The same is true for entire trains if metal wheels are used. The metal wheels on the last car will cause a short as they cross the gap in the tracks where polarity changes.

Turntables and wyes

The second most common issue comes when using turntables. The problem occurs when you rotate the turntable, which in effect reverses the polarity of the turntable tracks. Once the turntable crosses the point where polarity flips, if the locomotive wheels bridge the gaps, or both rails don't change simultaneously, a short will occur.

To fix this problem, you need a way to correct the polarity disparity, which, as shown in **4**, can be done with switches or special wiring. However, if you use one of the DCC auto-reversing circuits mentioned before, they will instantaneously detect and correct any differences in polarity.

The final problem involves the use of wyes. These track arrangements, shaped like a triangle, allow a locomotive or whole train to be driven onto one leg of the wye, and then backed out through another leg, then pull forward again on the third leg, reversing direction of travel in the process.

However, as with reverse loops, due to the combination of track and turnouts, you end up creating short circuits in the process, **5**. Like the other situations, special wiring and switches can correct the problem. However, forgetful operators sometimes neglect to flip these switches.

Again, by using DCC auto-reversing circuits wired in place of the switches, **6**, the polarity will be corrected automatically without your train crews having to remember to do anything but run their locomotives.

In general, auto-reversing circuits designed specifically for use with DCC are something of a cure-all for situations where previously one or more DPDT switches and extra wiring were required.

Various other versions of these circuits designed for automatically correcting frog polarity and lining turnouts are also available. For more information on this subject, I dedicated an entire chapter to it in my book *Wiring Your Model Railroad* (Kalmbach Books, 2015).

Avoiding short circuits with DCC

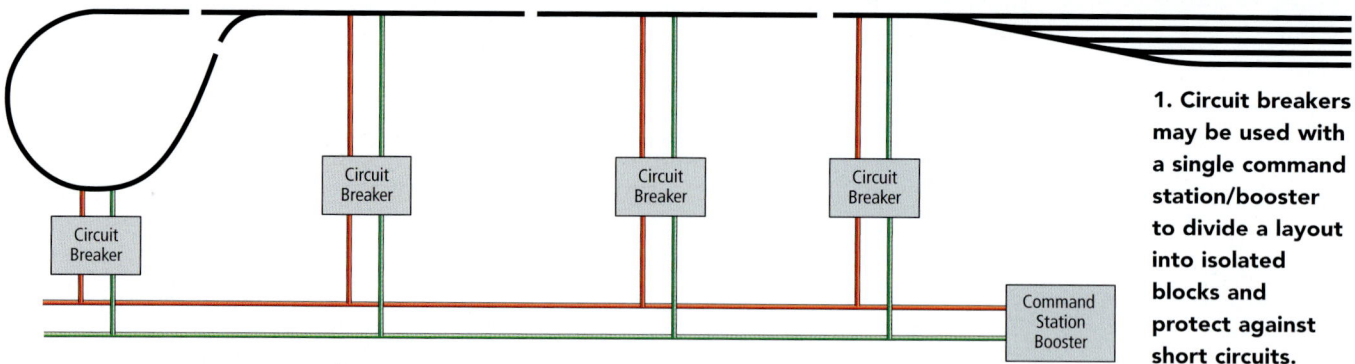

1. Circuit breakers may be used with a single command station/booster to divide a layout into isolated blocks and protect against short circuits.

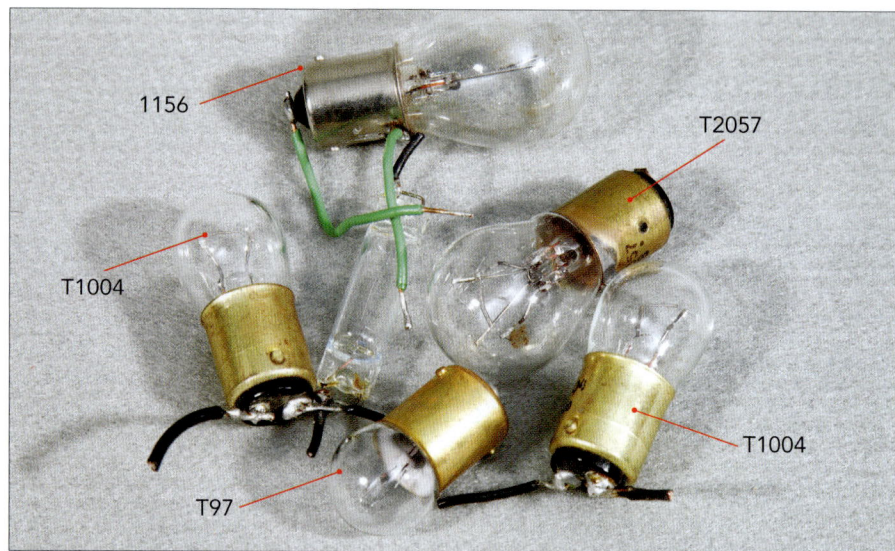

2. Many 12V automotive bulbs can be used as ballast lamps, but some experimentation may be necessary to find the optimal combinations of amperage. Larry Puckett found these bulbs, identified by number, at his local auto parts store. The packages were labeled with amperage and voltage ratings.

Dealing with short circuits has always been an issue in model railroading, and they became more of an issue when DCC came along. With DCC, all the trains on a layout may be powered by one command station/booster, and if it shuts down due to a short, *everything* comes to a halt. And with boosters supplying up to 10 amps, there's potentially a lot of wattage involved that can translate into heat under the right circumstances.

The most common causes of shorts are:

1. Tools, screws, and other metal objects left on the rails

2. Out-of-gauge wheels or oversized wheel treads running through turnouts

3. Derailments on turnouts at frogs and switch points

4. Running past rail gaps at the frog ends of power-routing turnouts

5. Polarity changes in turnout frogs, reverse loops, wyes, and on turntables

The first problem is easy to deal with—don't leave things on the rails. For the others, wheels should be checked when initially placed in service. Fixing issues with power-routing and all-live turnouts is covered on pages 30 and 32.

Dealing with polarity issues in item 5 can usually be handled with the various auto-reversing circuits from DCC Specialties, Digitrax, Model Rectifier Corp., Tam Valley Depot, and others. Another approach when using a Tortoise by Circuitron slow-motion motor on turnouts is to wire one of the two built-in single-pole double-throw (SPDT) switches to automatically correct polarity each time the route is changed (see page 26).

Our focus here is dealing with shorts that result in shutting down the entire layout. By dividing the layout into electrically isolated sections or blocks, **1**, we can prevent a short in one section from affecting the other sections of the layout.

In the early years of DCC, this meant installing a booster for each electrically isolated block in the layout, but that's expensive overkill. A less-expensive option is installing circuit breakers that work similarly to those in our homes. Circuit breakers for DCC range from simple and cheap to complex and expensive, and each has its own advantages. The least expensive option, which is actually a circuit limiter, is a simple 12V automotive light bulb, referred to in this use as a ballast lamp.

Ballast lamps

Ballast lamps, **2**, work by wiring a light bulb into one of the wires of the power bus powering the block. When a short occurs, the lamp turns on and limits the current in the affected block. One good thing is that as long as the ballast lamp is mounted in a visible location, you know where the short circuit is as soon as that light bulb turns on. As always, the devil's in the details, so let's take a closer look.

Ballast lamps work essentially as variable resistance electrical components. When the current passing through the filament is low, the resistance is also low and the bulb doesn't light up. However, when a short occurs, the full current of the booster attempts to flow through the bulb filament, the bulb lights up, and

3. The NCE Circuit Protector ballast lamp-based device provides protection for up to six power blocks at a rating of 1 amp each with the standard bulbs and 1.75 amps with optional bulbs.

4. Electronic circuit breakers are available from several manufacturers in various configurations and capabilities. This is a PSX-4 from DCC Specialties.

resistance increases.

The result is that the current flowing to the affected block is prevented from exceeding the maximum operating current of the bulb. The bulb most commonly used for this purpose is the no. 1156 taillight bulb, and its maximum operating current is about 2.1 amps.

The weak point of ballast lamps is they don't stop current to a short, they only limit it to the operating current of the bulb. That bulb can get hot, and so can the point where the short is occurring.

Another aspect of the variable resistance feature of ballast lamps is that, like any resistor placed in your track wiring, they will reduce the voltage, and the voltage will vary depending on how much current is flowing.

Normally, the current draw of a single locomotive isn't enough to significantly affect track voltage. However, when several locomotives are running, they can draw 1 amp or more. Then the bulb will start to glow, voltage will drop, and your trains will slow down. This will happen whether there's a short or not.

So with an 1156 ballast lamp, if you're running several locomotives pulling a total of 1 amp, you'd lose about 2.8V from your track voltage. Instead of, say, 14V on the track, you'd be left with only about 11.2V, and your locomotives would slow accordingly.

You could compensate by increasing the booster output voltage to 15.8V. However, the track voltage will still vary depending on how much current is flowing through each ballast lamp.

Another downside is that the 2.1 amps used by the 1156 during a short circuit is subtracted from your total available amperage from a given booster. So the 1.44 amps provided by the NCE PowerCab power supply would be inadequate and everything would come to a halt as soon as a short circuit occurred.

Even systems like a Digitrax Zephyr, with its 3 amp output, would only have .9 amp left when a short occurs. A short while running other trains could lower voltage enough to cause erratic operations with some sound decoders.

There are other bulbs available. The NCE Circuit Protector, **3**, comes with ballast lamps rated at 1 amp; bulbs rated at 1.7 amps are available as an option. For more information on ballast lamps, visit Allan Gartner's website (www.wiringfordcc.com) and Marcus Ammann's website (www.members.optusnet.com.au/nswmn/1156.htm#_top).

Circuit breakers

On the more expensive side are circuit breakers, **4**. These devices, available from DCC Specialties, Digitrax, MRC, NCE, Tam Valley Depot, and others, sense when a short circuit occurs in the block they're protecting and completely shut off power. Most are solid-state electronic devices, although some may use relays and be a bit slower as a result. Most of these circuit breakers can be set up to activate at various trip currents. They're also self-resetting and will repeatedly attempt to re-establish power.

There are a couple things to look out for with circuit breakers. First, low-output boosters such as the PowerCab may not be powerful enough to stay on if the circuit breaker cycles on and off during a full short. To counter this problem, some circuit breakers can be configured to remain off in case of a such a short, and then be restarted using a remote switch once the short circuit is removed.

Unlike ballast lamps, circuit breakers are wired into both wires of your power buses. Because they can be set for specific trip currents, they allow you to customize your power districts and manage the power available from your boosters and command station in much the same way the circuit breakers in your house circuits are rated for different amperages. This allows you to operate a large layout using fewer boosters.

Wiring all-live turnouts

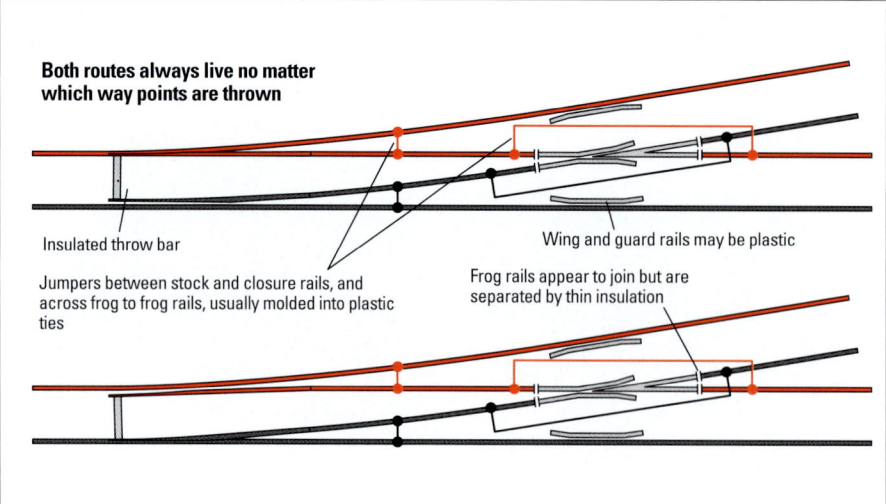

1. By providing direct connections between the stock and closure rails and isolating the frog, all-live turnouts avoid many of the issues found in power-routing turnouts and lessen the chance for short circuits.

All-live turnouts are designed so the point, closure, and frog rails on opposite sides of the turnout are all electrically independent, and the frog is electrically isolated, **1**. These features reduce the chances for short circuits, which is why all-live turnouts are generally considered to be DCC friendly.

Powering all-live turnouts in most cases is simply a matter of connecting the turnouts to a section of track that's already powered, by soldering them with rail joiners at the point end. You can also solder feeders to the stock rails and frog rails.

On most all-live turnouts, the frog rails are wired to the closure rails via jumpers embedded under the frog. However, on some others, like Micro Engineering turnouts, the frog rails are electrically isolated. Many all-live turnouts also have jumpers between the closure rails and their adjacent stock rails so their power isn't dependent on physical contact at the points.

The Peco Insulfrog turnout presents a special situation, since it has no jumpers between the closure rails and the stock rails, making it dependent on physical contact at the points for power. However, since its frog rails are continuous with the closure rails, soldering the frog rails to powered rails will feed power to the closure rails as well.

I suggest soldering jumpers between the closure rails and stock rails to make the turnout truly all live. However, the frog itself is plastic with metal wing rails, so it can't be powered. Peco tried to minimize electrical pickup issues by making the frog very short, therefore reducing the likelihood that locomotives will stall on it. However, if your locomotive wheels get dirty, then look out for erratic operation with unpowered frogs. Remember Murphy's Law: "Anything that can go wrong, will go wrong." So plan and wire for the worst-case scenario. The easiest way to do that is by using turnouts with frogs that can be powered.

There's another way you can use the Peco Insulfrog design to your advantage. On page 32 I explain and show how to use the power-routing feature of power-routing turnouts to energize a spur track. As built, the Peco Insulfrog picks up power at the points. Only one route at a time will be powered, assuming additional feeders and jumpers aren't installed.

So in effect it can be used as either a power-routing or all-live turnout depending on how it's installed and wired. Just connect the turnout to live rails at the point end and connect a spur to the diverging set of rails, and the rails will only be powered when the points are set for them. If you use this little trick, make sure to clean the turnout points regularly to prevent erratic performance.

Powering frogs

With the exception of Peco Insulfrog turnouts, which have plastic frogs, most other all-live turnouts have frogs that can be independently powered. But why power frogs at all? Since isolated frogs are electrically dead, some types of locomotives can lose power going over them and stall. This is usually only an issue with short-wheelbase locomotives.

Modern steam locomotive models typically are designed to pick up power from all their drivers, and in some cases, from other wheels as well (often the tender wheels). However, steam locomotives have long, stiff wheelbases over their drivers, and they can lift when crossing frogs if there are any irregularities in the rails.

This can result in loss of power if there are no additional power pickups. Diesel models aren't as sensitive to this issue, since their front and rear trucks can pivot independently.

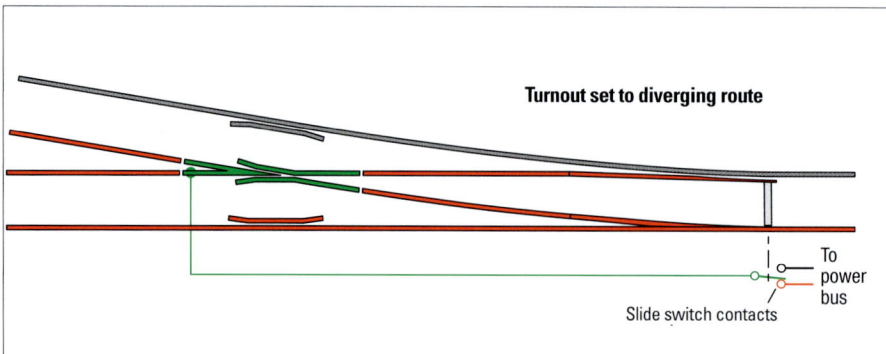

2. A simple single-pole double-throw (SPDT) switch provides the basis for powering isolated frogs.

30

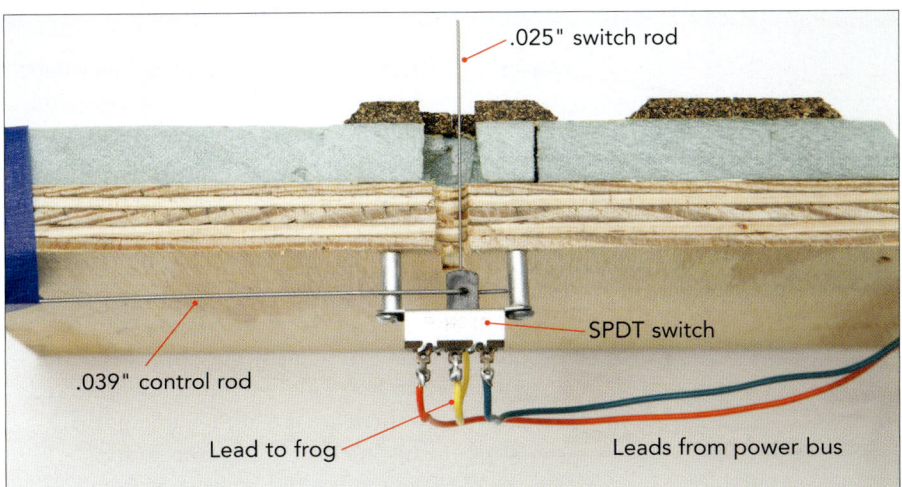

3. A single-pole double-throw (SPDT) or double-pole double-throw (DPDT) switch can be used to switch points and control frog polarity. It's connected to the turnout by the switch rod. The control rod leads to the fascia, where it's attached to a push/pull knob.

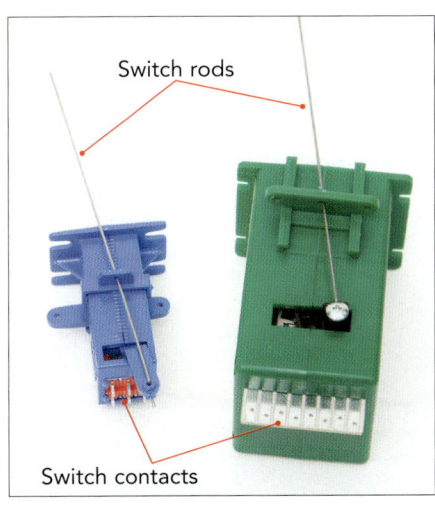

4. Blue Point switch machines, left, work like a manual version of the popular Tortoise by Circuitron switch motor.

Powering frogs can create an added expense depending on what you want to achieve in your operations. Push-button controls and electrically driven switch motors are going to be more expensive than manual methods.

No matter which method you use, most depend on a simple single-pole double-throw (SPDT) switch to change the polarity if the frog, **2**. With SPDT switches, two track bus wires are fed into the switch; the center pole is connected via a wire to the frog. When the slide switch is moved, it changes polarity automatically for you.

Turnouts made by Peco and Micro Engineering have springs built into their points that hold them in place when lined with a flick of a finger. However, this doesn't change frog polarity. One cheap and easy way to both switch the turnout points and power the frog is to use an SPDT slide switch installed under the turnout, **3**.

A stiff wire (.025"-diameter or larger) installed in a hole through the roadbed moves the points. Another wire (.039" or larger) running from the switch through the fascia allows you to use a push/pull knob to control the switch. Connect wires from the power bus to the outer contacts of the SPDT and run a wire from the center contact to the frog. Use an ohmmeter to be sure you have the switch oriented correctly in relation to the leads from the bus wires.

Every time you line the switch, the polarity will be changed for you. Caboose Industries makes ground throws with built-in SPDT switches that offer similar simplicity at a slightly higher cost than plain SPDT switches.

At the next level of expense is the Blue Point switch machine, **4**. This manually operated device resembles a Tortoise switch motor. It has a double-pole double-throw (DPDT) switch for controlling frog polarity, but you can use half of the DPDT switch, making it an SPDT, and wire it the same as described in **2**.

The Tortoise, **4**, is probably the best known of the stall-motor-type switch machines. It gives excellent slow-motion operation of turnouts and includes two SPDT switches—again, wire it like the rest. Similar stall-motor devices such as the Switch Tender, Switchmaster, and the new Rapido rotary switch machine offer similar capabilities and approaches.

Finally, should you choose to use a switch machine or a Caboose Industries ground throw without an SPDT switch or contacts, there is another option—the Frog Juicer from Tam Valley Depot, **5**. This device is powered by wiring it to the DCC track power bus. A third wire is connected to the frog. Any time a metal wheel bridges the gaps on either side of the

5. The Mono Frog Juicer from Tam Valley Depot automatically senses when frog polarity is incorrect and instantly corrects it. Think of it as a solid-state electronic SPDT switch.

frog, the Frog Juicer detects whether the polarity is correct and automatically switches it instantly. This device works so reliably and quickly that you don't even notice it's there.

For a look at the other major type of turnout—power routing—turn the page.

2 Wiring power-routing turnouts

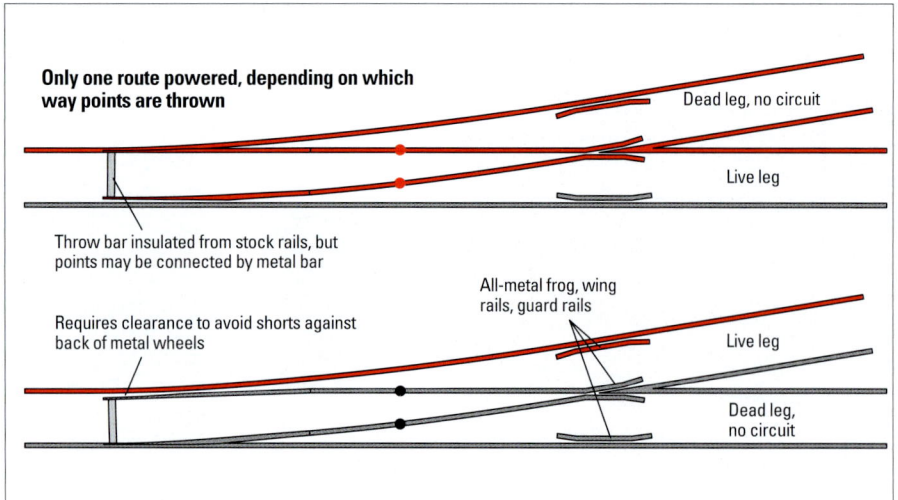

1. On a power-routing turnout, the point rails, closure rails, frog, and frog rails are always the same polarity. Only the route through the turnout set by the points is electrically live.

For decades, power-routing turnouts were the rule on model railroads. However, since the introduction of DCC in the 1990s, their popularity has steadily declined in favor of more DCC-friendly versions. So what is it about power-routing turnouts that led to their decline in popularity? And more importantly, how do you use them safely with DCC?

As their name implies, power-routing turnouts are designed to send power only to the route to which the points are set, **1**. They're also designed so that the frog has the correct polarity for the given route. Power is transferred from the stock rails to the point rails by physical contact, and from there transmitted to the frog and both frog rails by physical contact at joints and through internal wiring.

As a consequence of this design, both point rails, the frog, and both frog rails are always the same polarity. This can save a lot of wiring, eliminates unpowered frogs (improving operation), and explains their popularity—it can be used to your advantage or work against you.

The downside is that having both frog rails the same polarity can result in a short circuit if they're not electrically isolated from individually powered diverging rails. This is usually done by installing plastic rail joiners or leaving gaps at the ends of the frog rails.

The other problem is that if a metal wheel derails while passing through a turnout, it can create a short between the point and stock rails of opposite polarity leading to a dead short and the DCC booster shutting down.

Finally, if a locomotive's metal front wheels cross over the isolating gaps at the frog rails when the turnout is lined against the locomotive's route, it will create a short. This calls for extra attention on the part of operators and can lead to a lot of interruptions due to those booster shutdowns—this is why they are not considered DCC-friendly.

All-live turnouts, on the other hand, are often referred to as being DCC-friendly because they avoid the issues found in many power-routing turnouts. You can read why on pages 30-31. Their construction prevents the kinds of short-circuits that power-routing turnouts are subject to.

Practical uses

That said, there are applications where power-routing turnouts can be useful, even on a DCC-powered layout. One application I've seen on several model railroads uses the built-in power routing feature to park a locomotive on a spur and then shut off power by simply lining the turnout against it, **2**.

Using this method, power to the rails isn't actually cut off; instead, the two frog rails and any rails connected to them are set to the same polarity. This in effect kills power to the locomotive and prevents it from drifting off if the DCC throttle set to the locomotive's address isn't set all the way to "0" or the throttle gets bumped accidentally. It also means sound-equipped locomotives don't sit on sidings idling away during the whole operating session, which can get annoying after a while if there are lots of parked locomotives.

This approach can also be used for controlling the power to sidings and other tracks. However, there are some things to consider when also using feeders to power the siding tracks independent of the frog rails, **3**.

One great advantage to DCC is it allows you to operate multiple locomotives in the same area (or even on the same track) at the same time without having to electrically isolate each track. In order to do this, all tracks need to be powered with their own feeders instead of relying just on the frog rail connections that could interrupt power as the points are lined against a route. As you can see in **3** this can lead to shorts should the lead wheels of a locomotive bridge the gaps

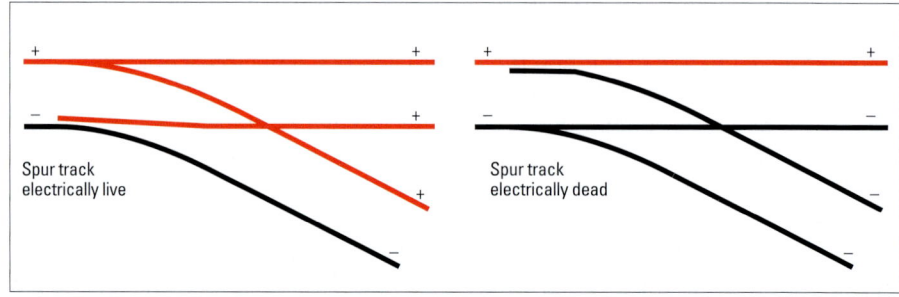

2. With the point rails set for the spur track, only those rails are of opposite polarity, allowing a locomotive to operate. With the points set for the mainline track, the spur rails are the same polarity and electrically dead.

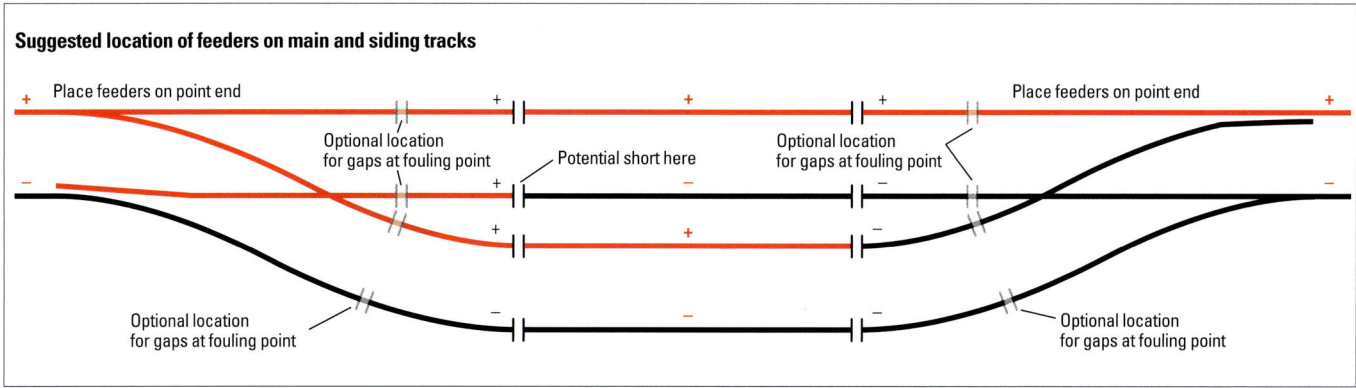

3. When using power routing turnouts to control a siding, Larry recommends putting gaps in the rails as shown, with feeder wires at the red "+" and "-", although you could skip the gaps on the outer rail of the siding.

where rails of opposite polarity abut.

This is less of a problem on long sidings than on short ones, since a locomotive has more room to operate without crossing the gaps. To limit this kind of short, the gaps at the frog rails should be set back as close to the frog as possible. I suggest putting them near the fouling point, since your engineers are more likely to stop before they get that close to the frog.

One of the few power-routing turnouts still in production is the Peco Electrofrog. These turnouts are well made and popular, so let's take a look at them and how they can be used with DCC.

The Electrofrog is designed as a true power-routing turnout with the point, closure, frog, and frog rails all receiving power through the physical contact of the points to the stock rails. When used as a power-routing turnout, if you connect the frog rails to powered closure rails, the frog rails must either be gapped at the connection or insulated rail joiners must be used to prevent shorts.

You also have the option of not connecting feeders to the diverging rails and powering them from the frog rails. In this configuration the diverging rails will only be powered when a route is set for them as in **2**. Be aware that if one route through the turnout is part of a main line that is powered elsewhere, that side of the frog must be insulated to prevent a potential short circuit.

HO scale Electrofrog turnouts can also be easily converted to an all-live

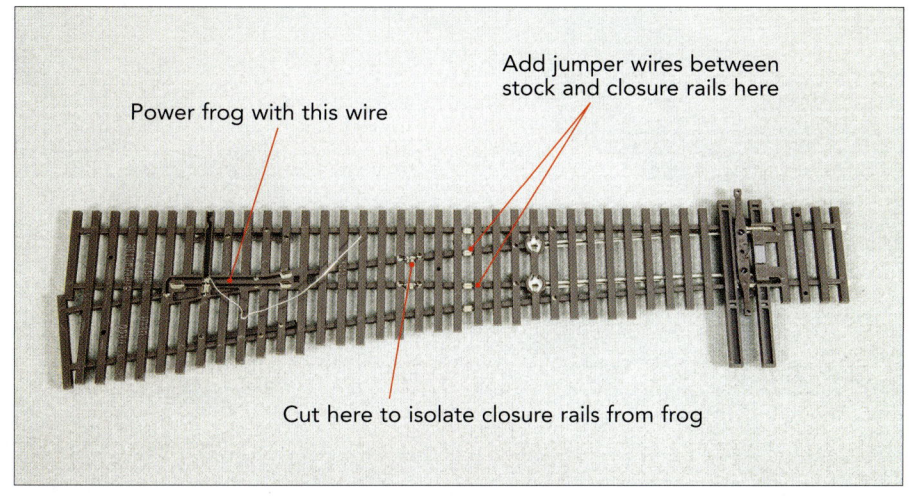

4. On the underside of an HO scale Electrofrog turnout, Peco has provided jumpers that can be cut to separate the frog and closure rails, locations where jumpers can be installed between closure rails and stock rails, and a wire to power the frog.

configuration with an independently powered frog, **4**. Converting an Electrofrog to a DCC-friendly all-live configuration can be done by snipping two small jumper wires between the frog and closure rails. This leaves a frog that can be independently powered by a single-pole double-throw (SPDT) switch such as those on a Tortoise by Circuitron switch motor—a wire already connected to the Peco frog is provided for this purpose.

I also suggest adding jumpers between the closure rails and the stock rails for reliable electrical connections, and isolating the closure rails from the frog itself using a cutoff disk in a motor tool. The closure rails can then be soldered to the stock rails on the connecting track to power them. These modifications may not be possible with Peco's and others N scale turnouts due to their smaller size and tighter tolerances.

What I've just shown you provides the basic details for using power-routing turnouts with DCC. More details on power-routing turnouts, along with other quick-fix methods for using them safely, are discussed in my book *Wiring Your Model Railroad*, available from Kalmbach Books (www.kalmbachhobbystore.com) and on my website (www.dccguy.com).

2 Electrical troubleshooting

If you have an electrical problem on your layout, working through a checklist of troubleshooting tips can help restore operations.
Photos by Bill Zuback

During my career as an electrical engineer, I've had to solve many electronics-related problems. For more than 50 years, I've been able to put my professional background to use in model railroading by developing electronic throttles, writing about direct radio control for scale locomotives, and authoring books on model railroad electronics and DCC, including *The DCC Guide: Second Edition* (Kalmbach Books, 2014).

One thing I've learned over the years is there's plenty of information on how things work, but little available on why things *don't* work. Remember, when troubleshooting electrical issues on a layout, keep safety first. Although most model railroad currents and voltages are relatively low, high voltage and current can cause a shock or worse.

Here are nine ways to diagnose electrical problems on a model railroad. Some are hands-on solutions, while others rely on the expertise of others. Using one or more of these approaches, you should be able to diagnose almost any problem on your layout.

by Don Fiehmann

1

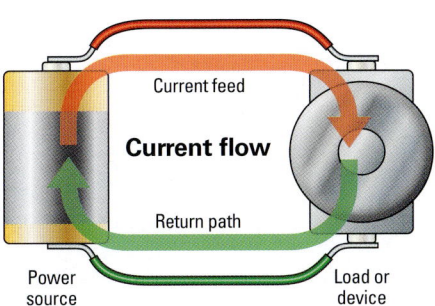

Think of a circuit as a circle. The circuit contains a power source, wiring and connections, and a load or device.
Rick Johnson illustration

Understanding basics

Troubleshooting starts with an understanding of electronics basics. There are three parts to a circuit: A power source, a power feed (wiring and connections), and a load or device. Any of these three parts can fail.

• Power source failures can be a dead battery or defective power supply. Recharge or replace the battery as appropriate, following the manufacturer recommendations.

If the power source has failed, contact the manufacturer for return or repair information. This is why it's important to fill out the warranty card.

If powered from a wall plug or power strip, check the switch. Also check fuses and circuit breakers. Sometimes a fuse can look good but be defective. To be sure, check the fuse with an ohmmeter.

• A broken wire or poor connection can cause a failure.

• The device itself can be defective. A burned-out lightbulb or bad motor can cause a failure.

An operating circuit is a closed circuit. An open circuit is one where there's a break in the circle. A short circuit is when the power flow takes a shortcut and fails to reach the output device.

2

Locomotives almost always stalled on this turnout on the Milwaukee, Racine & Troy. The turnout is mechanically sound, so the problem is electrical.

Electronic vs. mechanical problems

If a turnout isn't operating correctly it could either be a mechanical or electrical problem. A broken linkage is obvious, but if it's an electrical problem it will take more analysis.

Take the turnout shown above on the Milwaukee, Racine & Troy, Kalmbach's HO scale club layout. At casual glance the turnout looks fine. The switch rod is in good shape and travels through its range of motion smoothly. The points make even contact with the stock rails. There's no paint or scenery glue interfering with electrical conductivity. However, almost all DCC-equipped locomotives (except those equipped with TCS Keep-Alive or similar capacitors) stall on this turnout. Why? That requires a bit more digging.

We know the turnout works well mechanically, so we'll rule that out. Electrical problems are often hidden, and that's causing the problem here. It could be the phosphor bronze contact strip under the switch rod, a cold solder joint, or a bad electrical connection through the rail joiners.

3

If a model railroad isn't running properly, check the wiring. Look for damaged wires, loose connections, bare wires, or changes in how the wiring was installed.

Troubleshooting techniques

Here are a few thoughts that can help you find a problem.

• If you're having a hard time finding the problem, you may be looking in the wrong place. Think of all the possible causes of the electrical problem, big and small, and work through that checklist.

• Try checking the last place you worked on the layout before the trouble started. Did someone accidentally drive a staple through the wiring? Could a wire have been pulled loose? Did two bare wires (or joints) get pushed in contact with each other? Was there a change in wiring?

• Talk the problem over with someone else. Even if they don't understand the subject, just talking about it will keep you thinking about a solution.

• Sleep on it. You may be sleeping but your brain isn't. You may wake up with an epiphany at 2 a.m. However, it may be best to write down the idea and save it until morning. You may introduce more problems to your model railroad working on it in the wee hours of the morning.

• Expect the unexpected. Gee, how did that happen!

4

Books and magazine articles contain useful information on how electrical-related items should work on a model railroad.

Ways to find trouble

When trouble occurs on a model railroad, we start by thinking of all the complex things that could be wrong. Instead of over-analyzing, look for something simple, like a plug that got knocked out of the wall socket, a metal coupler height gauge left on the rails, or a stray uncoupling pin or metal detail part that fell across the rails somewhere.

If the simple solution doesn't solve the problem, use this as an opportunity to learn. Check magazine articles and books to figure out how it should work. *Model Railroader* has published many articles on electronics over the years. You can read all the electronics-related stories published in MR in the magazine's online archive at www.ModelRailroader.com/AllAccess. In addition, Kalmbach Books has titles on model railroad electronics and DCC.

If you can't find the answer in a book or magazine, go back to the product paperwork. Keep instructions, manuals, and exploded-view diagrams in a safe place, as these may be your only sources of information.

5

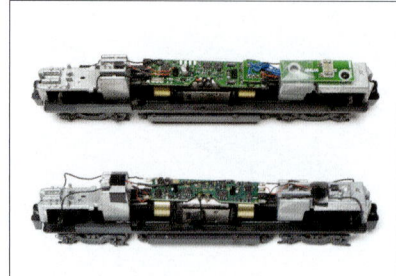

A whiff of smoke indicated the decoder on the Atlas HO model (top) was ruined. It was replaced with a TCS A6X board-style decoder. *Jim Forbes*

Analyze the situation

Here are a few things that can help you analyze the problem.

• Observation and smell. Is the power turned on? Did you smell anything or see smoke when the problem occurred? Are there any blackened or burned parts?

• Substitution. Replace elements of the system until you find the one that's causing the problem.

• The educated guess. You think you know what the problem is because you once had something similar occur.

• The crossed-fingers approach. You fix other problems hoping that it will also cure the latest problem.

• Divide and conquer. Start in the middle, then divide the wiring in half. Continue dividing until the problem is found.

• The scientific approach. Get serious and do some research. Dig out some test equipment to analyze the problem.

• Heat can cause things to fail. Electronic devices can return to normal when cool.

• If all else fails, read the instructions. They may contain troubleshooting tips.

6

A quarter across the rails is a way to test electrical connections. If wired properly, the power pack will detect an overload or the circuit breaker will trip.

Wiring and connections

About 90 percent of electrical problems are caused by poor connections. Wheel-rail contact is the first thing to check.

The next area to look at is layout wiring and connections. Look for broken wires or bad connections. A cold solder joint connection may look good but still not make a reliable connection.

Connections to the rails can be tested with a simple automotive lamp with a couple of wires connected. Or try the quarter trick, as shown in the photograph above. With the track power on, place a coin on the rails. The booster or circuit breaker should trip or the power pack should indicate an overload.

Rail joiners aren't always reliable and can fail to make connection, which is why some modelers solder joints. A feeder wire going to each section of rail is a more reliable solution.

Screw connections to power packs or boosters can loosen over time and need to be tightened. Insulation-displacement connectors (IDCs, also called "suitcase connectors") can fail, especially if the wrong size of connector was used for the wire size.

7

Before sending your model back to the factory, try resetting the decoder. Here, the decoder's configuration variable 8 is being set to "8."

Problems with devices and locomotives

When a locomotive stops working, go through this checklist:
• Is there an open circuit? If so, a quick push on the locomotive will determine if it was a contact problem.
• Is there power to the rails? If power to the rails is OK, try another engine.

If a device has voltage and still fails to work, be sure the return (ground) wire and connections are OK and the circuit is complete.

On decoder-equipped engines, test the headlight. Is CV19 set to "0"? Try short address 03.

If all else fails, reset the decoder back to its factory settings (CV8 to 8 on many decoders).

• Is there a short circuit? If it's a short circuit, check for a derailment or a turnout lined the wrong way. If lining the turnout doesn't fix the short, remove cars and locomotives from the track to see if that clears the short (metal wheels sometimes bridge gaps or touch multiple rails at turnouts).

• Did you test the decoder installation? If you install a DCC decoder or work on an engine with a decoder, always test it on the programming track before running it on the main line. If you put an engine on the main that has a short circuit, it may damage the decoder.

8

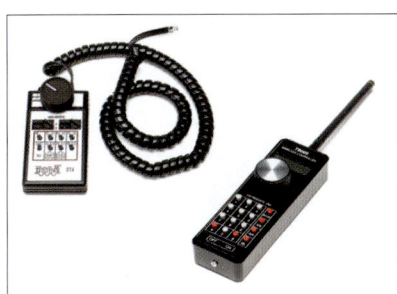

The tethered Digitrax UT4 throttle can be set up for infrared wireless operation (left), while the CVP9000E (right) uses a radio frequency.

Wireless hints

There are two types of wireless connections: infrared (IR) and radio frequency (RF). Infrared is like a TV remote. With IR, you need line-of-sight between the remote and the receiver. Sometimes the beam can be bounced off a wall. Radio frequency is much more reliable.

Working with radio waves turns out to be more of an art than a science. Here are a couple of suggestions when installing a wireless system.

• Radio waves don't propagate well through the human body. When installing the DCC transmitter, it should be located in front of the operator. Choose a high location like the ceiling.

• Concrete walls can also be a problem. It's not the concrete itself but any steel rebar in the concrete that can attenuate the signal.

Metal objects near the transmitter will also absorb some of the RF energy. Sometimes just moving a few feet one way or the other will improve the signal. If all else fails, try installing a second receiver.

9

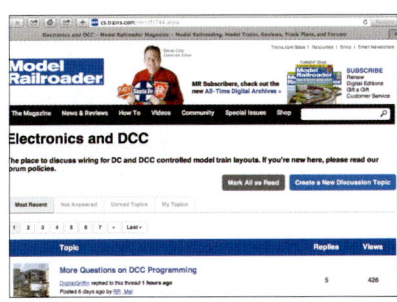

The Electronics and DCC section of the *Model Railroader* forums provides a way for hobbyists to find solutions to problems on their layouts.

Worldwide solution

The Internet is a great source for solutions for model railroad problems. The *Model Railroader* forums and DCC Corner columnist Larry Puckett's website (www.dccguy.com) are two places to start. In addition, Yahoo has many groups that can help with most questions. There are groups that focus strictly on wiring and DCC. Some groups specialize in one DCC manufacturer.

To find a group that matches your interest, do a search in Yahoo Groups (groups.yahoo.com) or in the newer www.groups.io. Most groups are open to new members, but some have restricted membership.

These groups span the world. I've seen someone with a problem in Denmark helped by a modeler in Australia. You can search past topics for previously discussed problems and answers.

Computer interface options for DCC

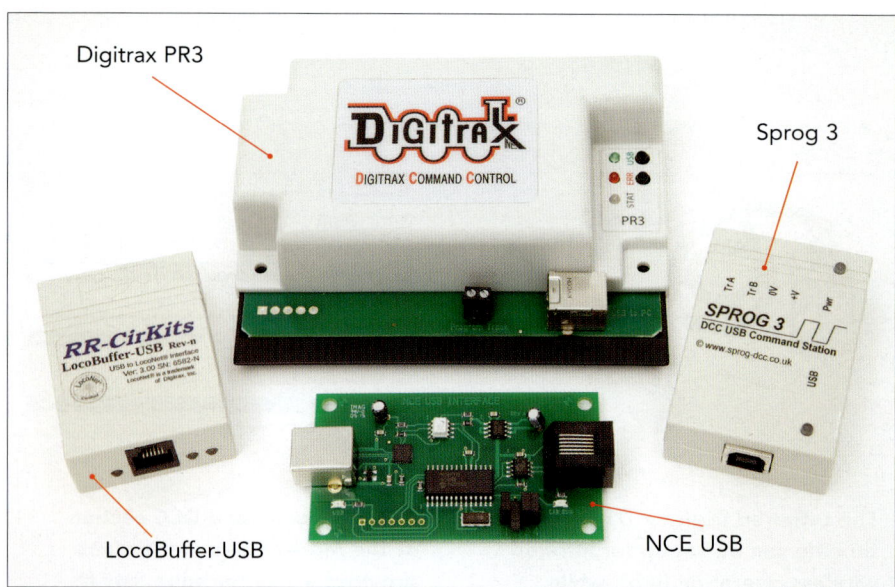

1. Many DCC manufacturers offer USB interfaces for connecting to a computer. Among the most popular are (clockwise from left) the LocoBuffer-USB, Digitrax PR3, Sprog 3, and NCE USB.

On page 80, I'll you how to program decoders using DecoderPro software. Let's take a look at three other computer interface options plus a bonus device, along with a couple other devices that can boost their performance.

These interfaces come in several types—those I'm aware of are system-specific. Some can be used as a stand-alone programmer, and at least one can double as a command station. There are also interfaces that can connect a command station to a computer for signaling and layout control.

If your system isn't among those I discuss, be aware most DCC manufacturers offer a similar interface for use with DecoderPro, and the Java Model Railroad Interface (JMRI) website (www.jmri.sourceforge.net) provides information on many of them on its hardware page.

RR-CirKits' LocoBuffer, **1**, has been around in various versions for well over a decade, and as a consequence has been one of the long-running favorites for use with DecoderPro. It functions solely as an interface between a computer and a Digitrax LocoNet, but not as a stand-alone programmer. Since it works only through a command station, it can be used for both service and operations (ops) mode programming. It's often paired with RR-CirKits' many other components and with JMRI for block occupancy detection, signaling, and layout control. The version I tested has a USB interface that operates off computer power.

The Digitrax PR3, **1**, was introduced in 2008 with a built-in USB interface. It can be used either as a stand-alone programmer or as a computer interface for programming and controlling a layout. When used with a command station it supports service and ops mode programming, and as a stand-alone programmer, it supports service mode programming. It was later released as the PR3 Extra, then the PR4 with a power supply. An interesting feature is it can be used with Digitrax's SoundLoader software to install sound projects in compatible Digitrax decoders.

Sprog 3, **1**, is an import from the United Kingdom that's increasing in popularity. This device is powered by a 14VDC, 3-amp power supply, contains a USB interface, and can be used as a stand-alone programmer for service mode programming. The Sprog 3 can also function as a DCC command station, which means you can also use ops mode programming. You can even use it to run a model railroad, as long as you're happy with DecoderPro's virtual throttle or the WiThrottle or Engine Driver apps for your smart phone.

NCE's USB interface, **1**, is designed specifically for use with its introductory DCC PowerCab system, **2**. The PowerCab includes a command station and booster in the handheld throttle, giving it the ability to program decoders in both service and ops mode. The USB interface allows the PowerCab to be used with a computer program like DecoderPro.

Beginning a little over 10 years ago, modelers started reporting issues with programming some sound decoders. The main issue was that some components in sound decoders drew enough current to interfere with the programming process. SoundTraxx responded with the PTB-100, and DCC Specialties released the PowerPax, **3**. Both function by supplying current to the programming track to charge up these power-hungry components before the sequence of programming commands are sent.

Installation and use

Installation is pretty straightforward. After the device is plugged into a computer using a standard USB cable, Windows will attempt to install a device driver for it. Most of these devices come with drivers on a CD, along with excellent instructions, and

2. The NCE PowerCab has a command station in the throttle that can be used with the NCE USB adapter for programming decoders in both service and operations mode.

I had no problems installing them. If you do get an error or failure to install, just reinstall and follow the instructions carefully, step by step.

Once installed it's a simple matter to open DecoderPro, edit your preferences, and proceed with programming (see page 80). One tip: Ensure that the default baud rate matches that given in the instructions. In many cases the default value will work, but it's good to confirm it.

Connecting the programming track boosters is even easier. In both cases, two wires are connected to the programming track connections on the command station, two other wires then go to the isolated programming track, and finally the appropriate power connections are made. The PowerPax comes with a power supply, while the PTB-100 can be powered using the same power supply that feeds your command station.

Once these are connected in line to the programming track, their operation should be essentially seamless. Blinking LEDs signal when a programming command is being sent, or if there is a fault on the track, such as a decoder with a short circuit.

As a warning, keep in mind that both programming track boosters are to be used only with service mode programming on an isolated programming track—do not connect them to the main track! Also, they can't be used with the NCE PowerCab system, nor should they be used with the Sprog 3 when using ops mode programming.

In operation, the true interfaces such as the LocoBuffer, PR3, and NCE USB pass the programming commands from the computer to the command station through the USB connection. Since these function with a command station, you have the option of using either the isolated service mode programming track or ops mode programming on the main track, but there are tradeoffs.

The advantage of service mode programming is you can read back the configuration variable (CV) settings in the decoder. However, track current is

3. When programming sound decoders, you may discover you need a programming track booster like the DCC Specialties PowerPax (top) or the SoundTraxx PTB-100 (bottom).

limited to about 250 milliamps, so you may need a programming track booster for sound decoders. On the other hand, most systems can't read CVs in ops mode, but with full power on the track, programming is very reliable. So even if programming doesn't work in service mode, it usually will work in ops mode.

For the stand-alone programmers, such as the PR3 or Sprog 3, you don't have to drag your command station along if you want to program decoders at your club or a friend's house. The NCE PowerCab and its USB interface are so portable that it too is convenient to use as a stand-alone programmer. Plus, if you have a DCC system that can't operate locomotives and program them at the same time, having a separate stand-alone programmer gives you the flexibility to do both jobs simultaneously.

How well do these devices work? As a test I tried programming a standard DCC decoder (Digitrax DN93FX) and two sound decoders with indexed CVs (LokSound Select and TCS WOWSound Diesel). With the LocoNet-based devices I used a Zephyr DCS50 command station, and where appropriate I also used a PTB-100 programming track booster. I've posted a full explanation of my tests and results on my website (www.dccguy.com).

As expected, none of the devices had any problems reading and programming the standard decoder, even without the PTB-100. With the sound decoders, I did run into some problems: Only the Sprog 3 and NCE Powercab USB could reliably program all the CVs in service mode. In most cases where I had problems it took a few attempts to successfully read or write CV settings. However, once the decoder address was programmed, I had no problems programming these decoders using ops mode. Keep in mind that if you have any problems programming individual CVs using your command station alone, you will not get better results when using DecoderPro, since it must work through your command station.

Which one is best for you? If you want an interface primarily for programming non-sound decoders and to connect your command station to a computer, go with the one designed for your DCC system. If you have any problems programming an occasional sound decoder, you should be able to at least program it using ops mode or with a programming track booster for service mode programming. For the most flexibility in reading and programming all types of decoders, then either the PowerCab USB or Sprog 3 seem to be the best choices.

2 Tips for better soldering

1. The Weller WLC100 soldering station can be adjusted over a range of 5 to 40 watts. Larry has since upgraded to a Hakko FX888D, which has a grounded tip, making it safer to use with electonic devices. Larry uses the brass turnings in a steel holder for cleaning tips. Pencil and flat screwdriver tips are available.

Brass tip cleaner
Moist sponge tip cleaner
Screwdriver tip
Pencil tip

Soldering is a skill every model railroader needs. When building a layout, soldered track feeders, rail joiners, and electrical connections are necessary for reliable DCC operations. In addition, most decoder installations require soldering at some point to add lights, speakers, connectors, and extra functions. Let's take a look at some of the tools that will make your job easier, followed by some tips on their use.

Iron and solder

The most obvious tool is a good soldering iron. Fortunately we have a wide array of electronic soldering irons to choose from, but how do you choose? As with most situations, you need to match the tool with the job. Small electrical components and delicate 28AWG wires call for small, low-wattage irons. However, soldering track feeders and rail joiners works best when you can put a lot of heat on the spot and get out before your ties turn into a pile of molten plastic or charred wood.

While you can purchase two or more irons of different wattages, I prefer a more flexible approach—an adjustable soldering station. The photo in **1** shows my Weller WLC100, but I have since upgraded to a Hakko FX888D. The Weller, although it worked well, had issues with leaking current at its tip, which could damage electronic components. The Hakko has a grounded tip which will not leak current. Like the Weller, the Hakko allows setting a specific temperature depending upon the type of solder and size of wire/connector being used.

Most soldering stations like these have interchangeable tips. I typically use a small pencil tip for delicate work and move up to a larger flat screwdriver tip for rail joiners and other projects requiring a lot of heat.

Solder varies in composition, with tin and lead in a 60:40 ratio commonly used for most model railroading purposes. However, 63:37 solder is also popular. An important distinction between these two formulations is that 60:40 solder has different temperatures at which it becomes a liquid and a solid, existing as a paste between these two points, whereas 63:37 solder passes from liquid to solid at 361 degrees F.

This is important, since if a 60:40 solder joint is moved while cooling, it might not form a solid connection, whereas the 63:37 solder solidifies immediately.

Hard solders, which often contain silver, are also available. They have higher melting points, near 900 degrees F. These are good for projects like building turnouts and adding details to brass locomotives. Some people also prefer lead-free solders.

You can find solder in strips, bars, and wire rolls of varying diameters; only the latter is suitable for most model railroading jobs, **2**. For years I've been using a 1 pound spool of Kester 60:40 rosin core solder with a diameter of .022". The small diameter makes it suitable for any job from soldering small electrical components to rail joiners and track feeders.

Because of the higher cost of solder containing silver (less than an ounce can cost as much as a full pound of 60:40 solder), I purchase silver solder as a small roll in a clear plastic dispenser.

Flux is an important component of any solder job. Flux prevents the formation of metal oxides during heating and allows the solder to make a solid, electrically conductive joint.

There are two basic kinds of flux: acid and non-acidic. Acid fluxes typically contain zinc chloride or other compounds that can leave corrosive residues that are difficult to remove.

These residues can corrode solder joints long after the job is completed and lead to failure months or even years in the future. That's why acid fluxes (or acid-core solder) should never be used in any electrical work or model railroading in general.

The flux I recommend for model railroading is a non-corrosive rosin flux. Rosin flux is available in both liquid and paste formulations and leaves

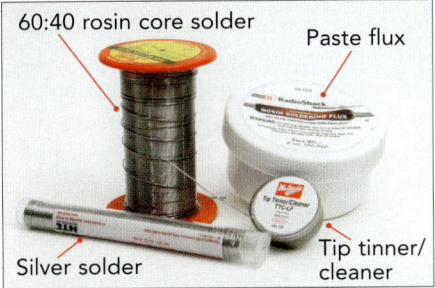

2. Larry's go-to solder is this 1-pound roll of .022" diameter tin/lead 60:40 rosin-core solder. The roll of silver bearing solder in the plastic tube is for special projects. Although the RadioShack rosin flux may be hard to find, similar products are available. The small container holds a mixture of rosin and solder powder for cleaning and tinning the tip in one step.

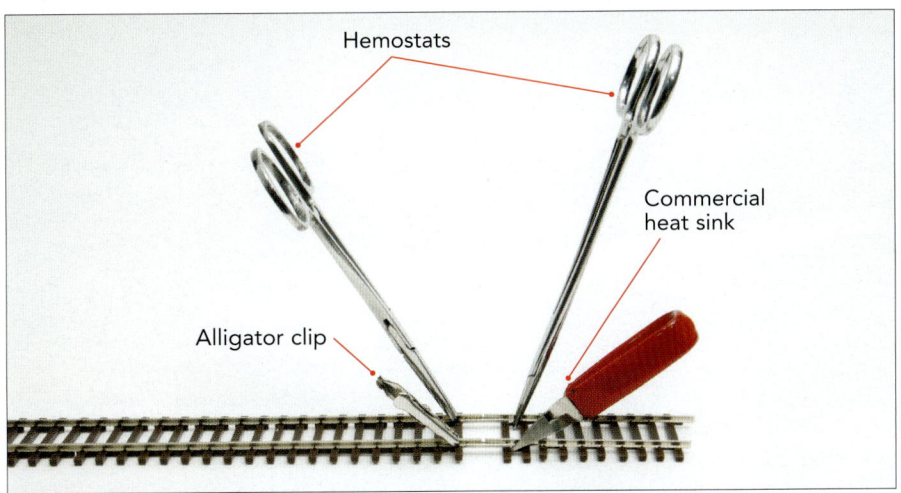

3. Heat sinks clip onto metal objects being soldered and prevent excess heat from melting or burning ties or damaging sensitive electrical components. Hemostats work well, but alligator clips or commercial heat sinks are smaller and are less likely to get in the way.

a non-corrosive residue that can be wiped away with alcohol. I prefer rosin paste flux, **2**, from RadioShack. This may be hard to find; however, similar products are available from electrical suppliers and on Amazon.com.

Working with fluxes is pretty straightforward, but you have to get the heat and solder applied before you burn off all the flux. That's why it's important to use a soldering iron big enough to heat the work quickly and evenly without overheating the flux.

If you burn off the flux, then the metal is likely to oxidize before the solder joint is complete and the joint will be weak and prone to failure. This is one reason many people prefer rosin core solder, with the flux making up 1 to 3 percent of the wire. I also apply a small amount of paste solder on large joints as well.

To get a good solder joint, it's important to keep the soldering iron tip clean and tinned. The key to remember is never use sandpaper or any other abrasive to clean the tip. Soldering iron tips have a copper core covered with iron. If you sand off the iron coating, the tip will be ruined. Weller makes a cleaning bar specifically for cleaning really dirty tips, but if you use the right solder, flux, and regular cleaning, you should never need one.

For years my common cleaning practice was to keep a moist sponge next to the soldering iron and wipe off any excess solder or residue on the sponge between uses. The WLC100 comes with a built-in sponge holder, **1**. Keep the sponge moist, not sopping wet, and the tip will come clean with a quick swipe.

More recently, however, I've been using a tip cleaner made of brass turnings in a metal holder, **1**. A quick poke of the tip into the brass turnings will remove any excess solder and residues, and you don't have to worry about keeping a sponge moist.

Tinning is the process of applying a small amount of solder to the soldering iron tip. This makes it easier to apply heat quickly to the components being joined. It's a good idea to always tin the tip before turning the soldering iron off. The solder protects the tip from oxidation and will prolong its useful life.

Weller and a number of other companies sell a tip cleaner/tinning paste comprising mainly flux with solder powder in it. By plunging your hot soldering iron tip into the paste, the tip is cleaned and tinned.

Avoiding damage to surrounding materials is another concern when soldering near plastic ties or anything else that can be burned, melted, or distorted. While it's important to quickly apply a lot of heat to a joint, metal components will also conduct that heat to other areas, which can result in damage.

To avoid this problem I always use heat sinks to isolate the area to be soldered. A heat sink can be something as simple as a piece of moist cotton cloth laid over the rails. Excess heat will be consumed by evaporating the water in the cloth, preventing damage elsewhere.

You can also use hemostats clamped on either side of the joint, **3**. A couple of small alligator clips or commercial heat sinks will likewise effectively intercept excess heat and are less likely to get in the way while soldering.

For most solder joints, I apply a small amount of paste flux using a small brush and quickly heat the joint with the correct iron tip. I regularly test the joint by touching the end of a piece of solder wire to it. As soon as the solder melts and starts to flow, I quickly apply as much solder as I feel necessary, then keep the work perfectly still while the solder cools. If you move the work before the solder solidifies, you may get a weak joint.

One sign of a good joint is a shiny surface appearance. A dull appearance indicates a weak joint. This can be corrected by quickly reheating the solder until it reflows, then holding it perfectly still. A quick scrub with alcohol on a stiff brush afterward will remove any rosin flux residue.

3 Are your locomotives DCC friendly?

1. Athearn's HO scale Ready-to-Roll SD40-2 in Southern's tuxedo paint scheme is one of Larry Puckett's favorite modern-era locomotives, and it's "DCC friendly"—a phrase that can mean several things, as Larry explains here.

It's still fairly common to hear models referred to as being DCC ready, **1**, DCC compatible, DCC friendly, DCC compliant, and it may have a National Model Railroad Association (NMRA) conformance warrant, but what's the difference? Well, DCC ready, compatible, and compliant generally mean the same thing, which actually can mean anything!

The model may have a socket for a decoder. The motor may be isolated from the frame. There may be space provided for installation of a decoder and speaker. Basically, when you see these terms, you need to do your research to see what you're buying.

The term "DCC ready" usually means the model has an 8-pin, 9-pin JST, or 21-pin socket that a decoder can be plugged into, **2**. Being DCC compliant and compatible usually mean the manufacturer says the model complies with the relevant NMRA standards and recommended practices (RPs). The devil in the details is what the manufacturer considers relevant.

Moving up to having an NMRA conformance warrant means the product has been tested, and says the model does in fact meet all the applicable NMRA standards and recommended practices (RPs). However, even this isn't a carved-in-stone guarantee, as I know of at least two conformance warrants that were suspended after it was found that the model didn't, in fact, meet all the standards and RPs. In one case the 8-pin socket was found to be wired incorrectly, and in the other case the motor was wired backward.

Now that manufacturers have had more than 20 years of experience with meeting NMRA DCC standards and RPs, it's rare to see any models that don't meet the minimum requirements for what I consider DCC-friendly status. However, there are tens of thousands of older models out there with motors that are not isolated from the frame, don't have a decoder socket, and don't have isolated coupler mounts. So why are these design issues a problem for DCC?

Challenges

One common instruction in decoder installation manuals is to isolate the motor—but what does that mean and why? Through the 1990s, many model locomotives were built with one pole of the motor connected directly to the frame, which also served as one leg of the electrical path.

This arrangement can all too easily destroy a newly installed decoder, because one of the decoder's motor output leads and one of its rail pickup input leads would be wired together through the motor. As the SoundTraxx manual says, "Failure to properly isolate the motor will damage your decoder and turn it into an effective, but short-lived, smoke generator!"

It's usually fairly easy to isolate the motor. If the motor is sitting directly on the frame, then a piece of electrical or Kapton tape placed between it and the frame will do the job. If it's held in place with a metal screw, make sure to replace it with a plastic or nylon screw.

Older Athearn-type model designs require a slightly different approach because the lower motor brush clip has two prongs that rub against the frame, **3**. These can either be removed or flattened, and electrical or Kapton tape installed as an additional check against shorting to the frame. In both cases the lead from the decoder can then be safely connected to the contact on the motor. Page 44 shows a decoder installation in this type of model.

Having an NMRA compliant DCC socket is an important feature in a model. For one thing, it implies the model is already wired for the most common functions, and it'll save you the extra work required to solder the wires to the motor, track pickups, and lights. It really means the model is plug-and-play unless sound is involved, in which case you'll have to add a speaker with the associated wire, heat-shrink tubing, and solder.

However, some decoders may not be plug-and-play ready. For example, the new SoundTraxx Tsunami2 TSU-1100 only comes with bare wires, not a plug, so you'll need to either hard-wire it or add a plug compatible with the locomotive.

One option with decoders like this is to purchase a decoder harness with the necessary plug from a DCC supplier, cut off the unnecessary end, and solder the wires from the plug to the wires of your decoder. Just make sure to match the wire colors and protect the joints with heat-shrink tubing. You can also purchase an 8-pin plug and solder it to your decoder wires, but that procedure requires a small-tipped soldering iron and a very steady hand.

Coupler mounts on most models made today are either cast into the body shell or use plastic draft-gear boxes. However, some older models had coupler boxes cast into the metal frame. For example, all my Atlas S-2 and S-4 Alco switchers have coupler boxes cast into the frame.

So what are the issues with cast-in coupler boxes? First, if you install all-metal Kadee couplers, it can lead to short circuits in any of these locomotives where the frame is one leg of the electrical path. Let's assume you've installed metal couplers in two old Athearn locomotives, and they're both facing forward. The frames on both locomotives being part of the electrical path will have the same polarity, since they'll both be electrically connected to the right hand rail.

Now let's say you reverse one of the locomotives so one is facing forward and the other is facing the opposite direction. Now one frame will be electrically connected to the right hand rail while the other will be connected to the left hand rail, creating a potential dead short through the couplers. This is a problem for both direct current and DCC users.

I say potential dead short because when new, the couplers and some frames may have enough paint on them to insulate them. However, over time the paint will wear off in spots, creating bare metal-to-metal contacts, which can create the short circuit. This type of short is difficult to find because the locomotives may not always be facing opposite directions, and the worn spots on the couplers and frames may not always line up.

The fix for this situation is to first be aware of the potential problem and to use plastic couplers. Kadee makes its series 20 couplers with plastic shafts and metal knuckles, which are perfect for this type of installation.

The addition of a lead from the truck frame to the motor on older Athearn models as I described back on page 25 doesn't resolve this problem because the metal truck frame is still conducting electricity to the locomotive frame, even though we've created a new, more efficient path for power pickup.

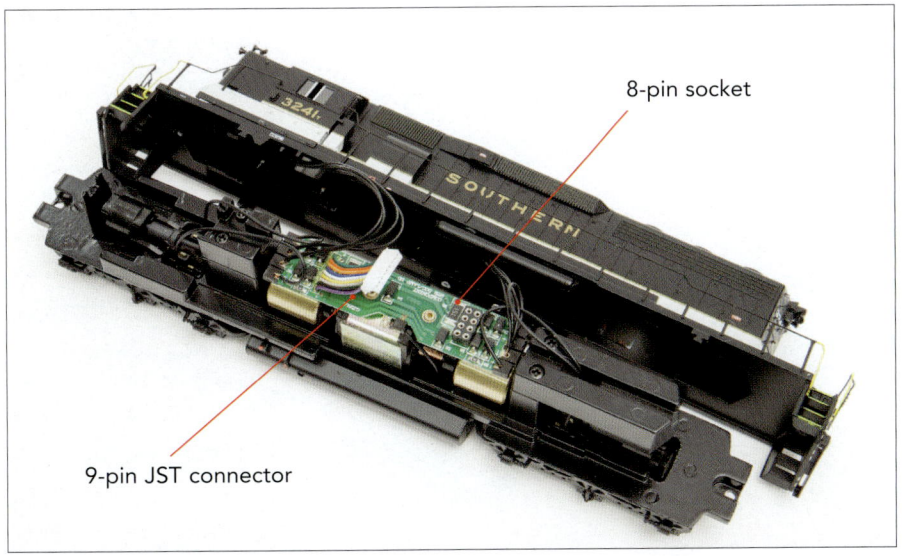

2. The 8-pin socket and 9-pin JST connector on Athearn's Ready-to-Roll locomotives, like this SD40-2, provide DCC-friendly plug-and-play compatibility with many decoders.

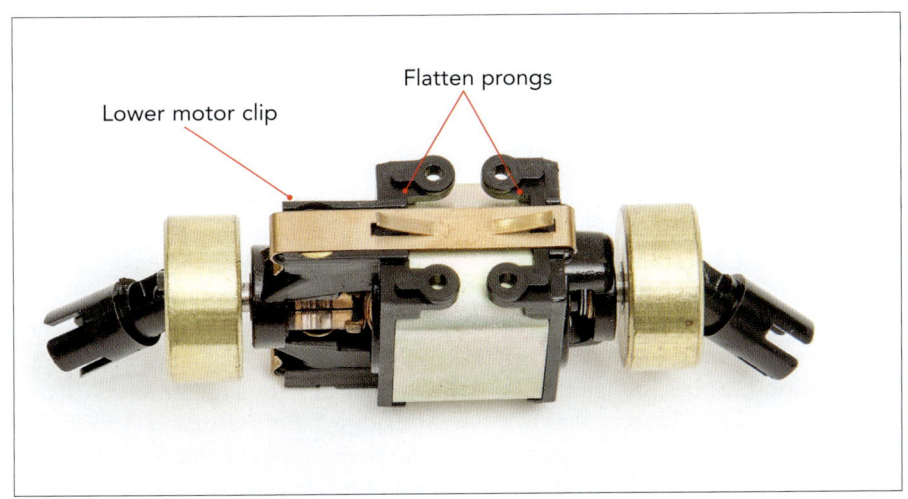

3. The two prongs on the lower motor brush clip of older Athearn locomotives connect the motor to the frame electrically. These must be flattened or removed, then insulated, to isolate the motor from the frame, which is part of the power circuit.

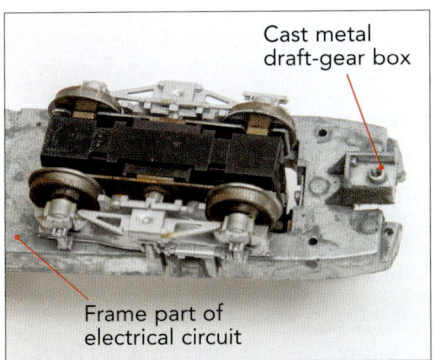

4. Draft-gear boxes cast into the frame of some locomotives can lead to a short circuit between locomotives if metal couplers are used.

3 Installing a basic decoder in a classic model

1. Amtrak General Electric AMD-103 no. 824 heads north through Rockfish, Va., on Larry Puckett's HO layout. Larry added a motor decoder and lights to this classic Athearn "blue-box" locomotive.

Sound decoders have become increasingly popular, but I know there are still a lot of model railroaders who simply aren't interested in them. Let's go through a basic installation of a non-sound DCC decoder, the NCE D13NHJ (and its No Halt capacitor unit), in an Athearn HO scale GE AMD-103, **1**.

As I've mentioned previously, the old Athearn "blue box" design that dates back to the 1950s served as the basis for many other manufacturers' model locomotives. For example, I often replace drive-train components in my Life-Like Proto 2000 locomotives with Athearn parts. Consequently, there are tens, if not hundreds, of thousands of models based on this design. This installation will work for many of them.

The AMD-103 model has provisions for twin headlights and ditch lights up front, along with a pair of lights in the rear bulkhead over the door. Fortunately, Miniatronics Micro Miniature 1.2mm diameter, 1.5V bulbs fit in the headlight holes. I sealed them in place with Testor's Clear Parts Cement. The ditch lights, being slightly larger, required Micro Miniature 1.7mm diameter, 1.5V bulbs.

I soldered an 820Ω resistor to one wire on each of the smaller bulbs and a 330Ω resistor to one wire for the larger ditch light bulbs. The ditch lights required different resistors because they drew a higher amperage. The instructions with the NCE decoder have an excellent table of resistor values for various combinations of bulb amperage and track voltage.

The resistors are soldered to the leads of the correct function wire—white for the front light, yellow for the rear light, and green and violet for the two ditch lights. Then I soldered all the other wires to the blue common wire. Be careful to keep track of all these wires and connect them to the correct colored wire in the harness, shown in **2**.

I don't recommend operating multiple bulbs off the same resistor, because if one bulb fails, the others will receive too much current and also fail. Likewise, I don't wire them in series because it's difficult to tell which bulb has failed.

I connected the wiring harness to the decoder and attached it and the No Halt capacitor unit to the shell using double-sided foam tape. I gathered the loose wires and taped them in out-of-the-way locations inside the shell, **3**.

Motor wiring

I removed the metal clip that runs from the top of the motor to the truck gear towers. This clip, shown on page 24, carries current from the right rail to the motor.

Next, I gently pushed against the vinyl rubber motor mounts using a small flathead screwdriver while wiggling the motor itself to ease it out of the frame. If you look at the bottom of the motor, you'll see a bronze clip with two small prongs sticking out. You can see an example of one on page 42. These prongs connect the motor to the frame completing the electrical circuit. Disabling them is essential to isolating the motor.

You have two options: Flatten the prongs using a pair of pliers, or just break them off. I usually just break them off. I removed the bronze clips from the top and bottom of the motor, being careful not to lose the springs that hold the motor brushes in place.

Using a TCS 6-wire harness (a 4-wire harness would work too, but I had the 6-wire harness on hand), I soldered a wire to each of the bronze clips and reinstalled them on the motor.

Now let's deal with getting power from the trucks to the decoder. First I soldered a wire from the harness directly to the top of each gear tower. Once this was done, I carefully removed the sideframe from the left side of each truck and sanded a clean spot on the side of the metal truck frame.

Next I removed any grease and

Fig. 2 Wire harness color codes
The functions of the wires on a Digital Command Control decoder are standardized with corresponding wire colors.

Wire color	Function
Green	Output 3 (F1)
Red	Right rail
Orange	Motor "+"
Blue	Common
White	Headlight (F0 Fwd)
Yellow	Rear light (F0 Rev)
Gray	Motor "–"
Black	Left rail
Violet	Output 4 (F2)

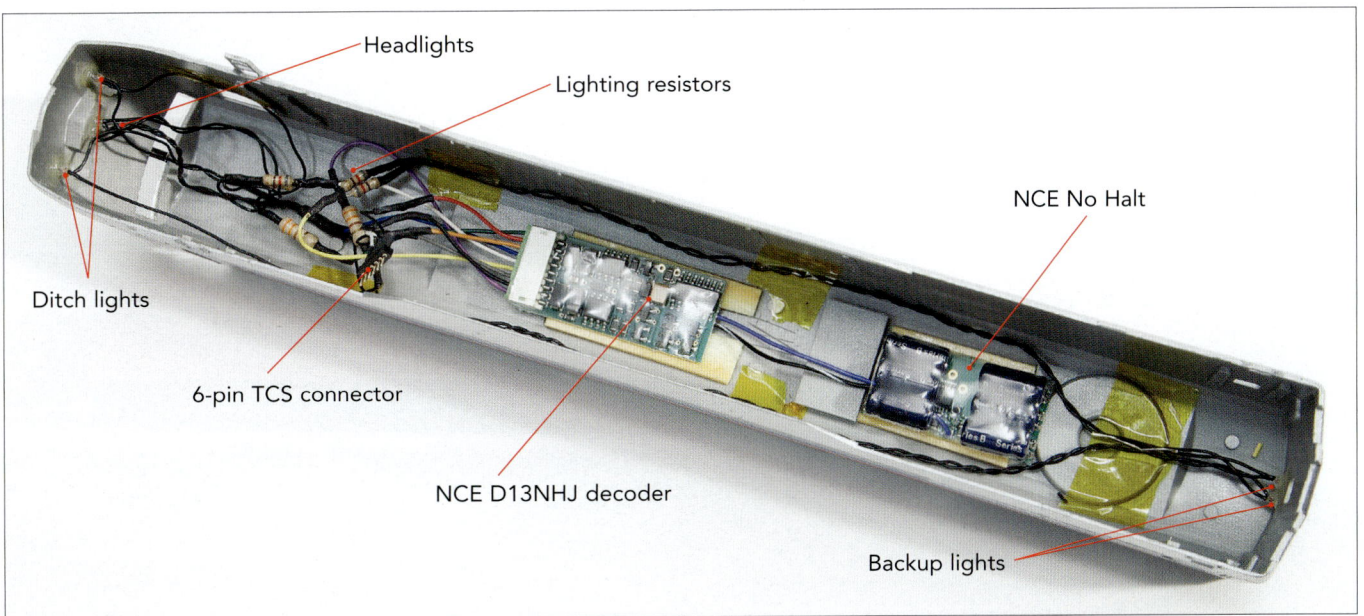

3. Larry glued the light bulbs into the model's headlight and ditch light openings. Then he soldered all the leads and resistors to the wiring harness, and installed the NCE decoder and No Halt capacitor unit.

oil with some alcohol on a cotton swab. Working as quickly as possible to reduce the chance of melting the plastic components on the trucks, I soldered another wire to each truck frame. Using rosin flux and pre-tinning both the frame and the wire ends made for faster work.

At this point I had six wires, one from each motor clip, and one from the opposite sides of each truck, connected to the wiring harness. If you're using a 4-wire harness, you can splice the truck wires from the same side of the locomotive together before connecting them to the harness.

With all the motor and truck wires connected to one end of the harness, I connected the male and female ends of the harness and soldered the appropriate wires on the free end to the matching decoder wires, **4**. Again refer to the table in **2**. In all cases I protected the joints with heat-shrink tubing.

Programming

I put the shell on the chassis, installed plastic couplers to prevent short circuits, and moved the locomotive to a service-mode programming track. I did a test read of the address to make sure there were no shorts.

Proceeding with the programing, I set a 4-digit address in place of the

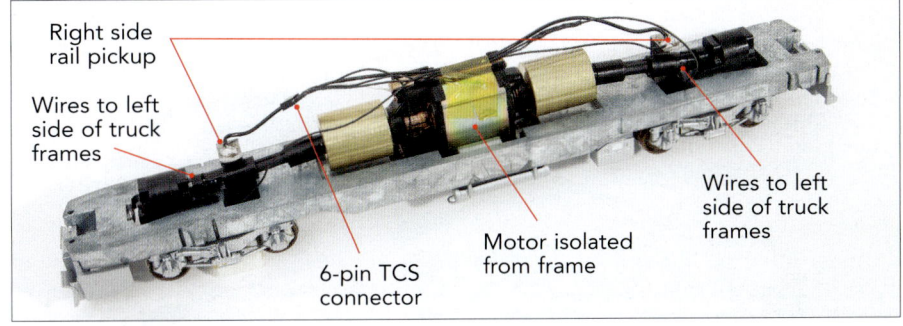

4. Larry soldered wires to the gear towers and to the left sides of the trucks to provide reliable electrical connections to the decoder.

default value of 3, then customized the light functions. The NCE instructions provided excellent examples for programming special effects such as Rule 17 dimming and ditch light operations.

I configured the headlight to operate normally in the forward direction and optionally use function 4 to dim it, by setting configuration variable (CV) 120 to a value of 32. To have the reverse light come on in reverse and go off in forward, I set CV121 to 2.

NCE offers two types of ditch light effects. With type one, the ditch lights are on whenever the headlight (F0) is on and flash when F2 is pressed. With type two, the lights are off until the F2 (horn) button is pressed, at which time they flash.

In all the photos I've seen of these locomotives, the headlight and ditch lights were on, so I used the type one option. This requires more complex programming. First, I set CV122 and CV123 to values of 56 and 60, respectively, for type one ditch lights. Then I set CV35 to 0 so output 3 wouldn't be controlled by F1, and CV36 to 12 to map outputs 3 and 4 to F2. Finally, I set CV118 to 20 so the ditch lights would flash for 5 seconds after the F2 button is pushed. Fortunately, all of these programming steps were provided in the NCE instructions, making this a quick and easy process, and the functions all worked correctly on the first try.

The classic model now has an upgraded look, realistic lighting, and a modern decoder.

3 Sound and DCC for the HO Atlas U30B

1. *Model Railroader's* HO scale Eagle Mountain RR project layout needed locomotives. Standing in for the full-sized railroad's U30Cs are Atlas Master Line U30Bs. After Cody Grivno gave them Kaiser Steel paint jobs, Eric White installed a SoundTraxx Tsunami sound decoder in no. 1038 for use on the layout. *Bill Zuback photos*

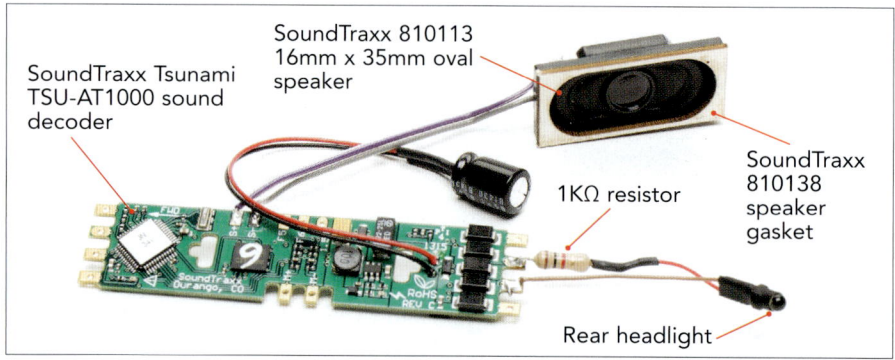

2. Eric used a selection of parts we had in the office to outfit the Atlas U30B with sound. In addition to the speaker, the rear headlight was soldered to the SoundTraxx Tsunami decoder.

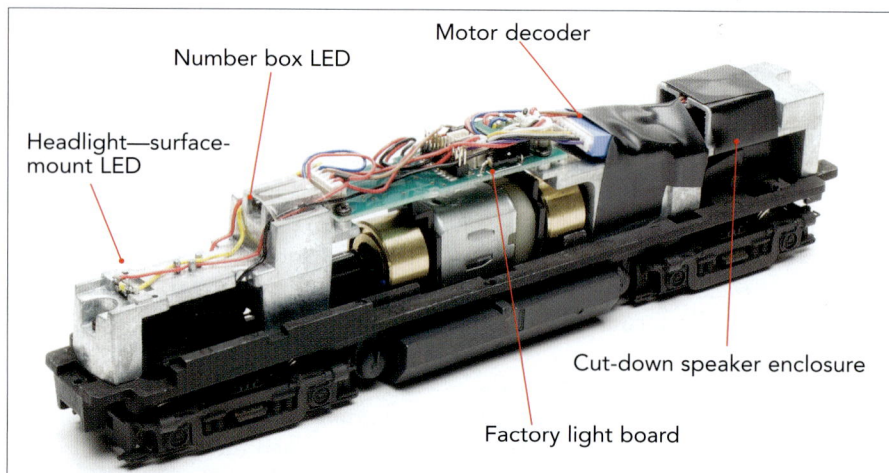

3. The non-sound decoder installation retained the original factory light board, but lost half of the speaker enclosure. Eric traded a functional speaker enclosure for the light board in the sound installation.

For *Model Railroader's* 2016 HO scale Eagle Mountain project railroad, we needed two General Electric U30B locomotives, **1**. Cody Grivno searched the internet and came up with a pair of older undecorated Atlas Master Line locomotives. Neither had a decoder, and we wanted to update both for DCC and equip at least one model with sound.

We had a SoundTraxx Tsunami AT1000 decoder on hand, which is intended to be a drop-in, board-replacement decoder for Atlas diesel locomotives. But our models were earlier production versions than what the decoder was designed for. We also had a 16 x 35mm oval speaker that would fit the locomotive and a gasket set for the speaker, **2**.

Cody painted and decaled both locomotives (see the December 2016 *Model Railroader*) and installed a motor decoder in locomotive no. 1037. While he was painting the other shell, I took the painted chassis and installed the sound decoder in no. 1038.

The model had a plastic enclosure to house the speaker above the rear truck and a printed-circuit (PC) board with an 8-pin DCC socket. But with the stock PC board in place, there wasn't enough room for the decoder and the

by Eric White

PC board, so the stock board would have to go.

I could've ditched the decoder I had and purchased a Tsunami with an 8-pin plug, such as a TSU-1000, but sometimes you need to go with what you have. Cody used an 8-pin jumper plug with a 9-pin socket to install the motor decoder in 1037, **3**, and he had to remove half of the speaker enclosure to get it to fit. Replacing the stock PC board looked like the best way to go.

Out with the old

The first step was removing the stock PC board. That left me with two metal tabs that could support the Tsunami. Since I didn't want the decoder to short out on the metal tabs, I cut two pieces of scrap plastic to fit between the mounting bosses on the tabs, then wrapped everything in non-conducting Kapton tape for insulation, shown in **4**. The pieces of plastic would later serve as mounting pads for the decoder.

Next was the speaker enclosure. The size was correct, but Atlas molded a groove in the top for the rear headlight and its wires. This groove made the box too shallow to accept the speaker I had. I sliced away the groove, then cemented .010"-thick scraps of styrene inside to seal the enclosure, **5**.

This left enough room for the speaker's magnet and the frame. I soldered wires to the pads on the speaker, then seated it in the enclosure. Finally, I added a stick-on gasket to eliminate any rattling between the speaker and the locomotive frame.

I test-fit the enclosure to the frame and found the screws that held it on were now too short with the additional thickness of the gasket. Since I didn't think I'd find the proper metric threads among our assortment of screws in the workshop, I used a drill the size of the screw head to countersink the screw into the housing. That was enough to get a couple turns of the screw into the frame. I took everything back apart and turned my attention to the lights.

The headlights on this model are LEDs, and I couldn't find any information about which wire was positive and which was the ground.

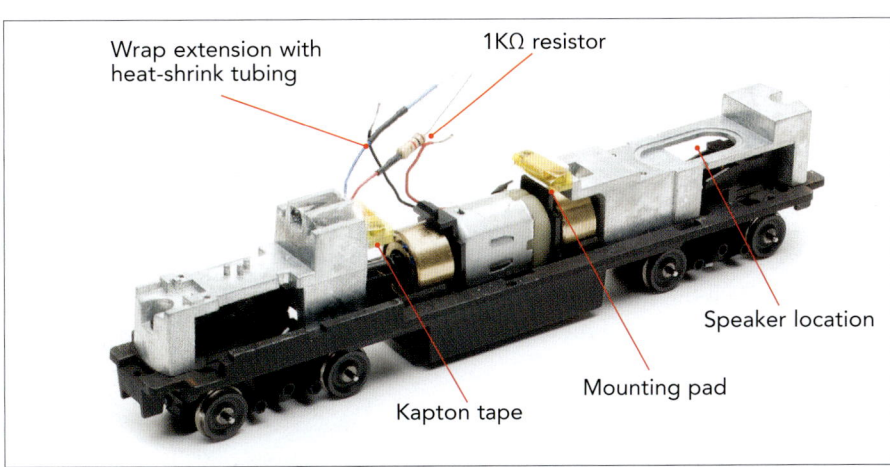

4. The chassis of no. 1038 was stripped of its factory light board and the mounting bosses are insulated with Kapton tape. An extension was soldered to the negative lead of the number box LED and a 1KΩ resistor was added to the positive lead.

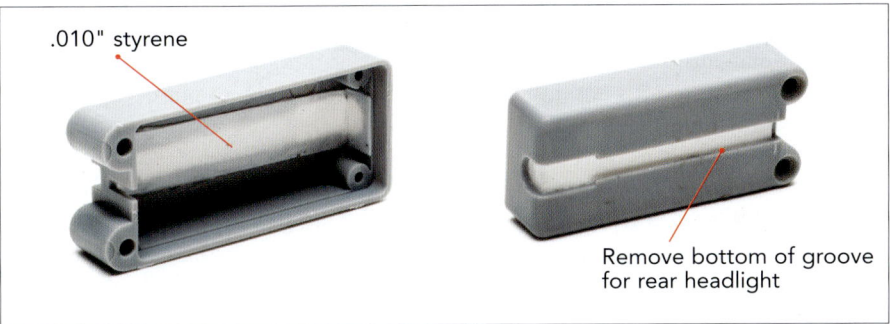

5. The sound enclosure in the Atlas locomotive was almost the right size for the SoundTraxx 16mm x 35mm speaker, but a groove in the base to accommodate the rear headlight and wiring made the enclosure too shallow. Eric cut out the groove and sealed the opening with thinner .010" styrene sheet.

This is important because LEDs will only light when wired in one direction.

There are three LEDs in the engine: a nose-mounted headlight, a cab-mounted headlight that also illuminates the number boxes, and a rear headlight that also illuminates the rear number boxes. All three lights had one red wire, and the other wire was either brown (rear), yellow (nose), or blue (cab). I figured the red wire should be positive, as most circuits have red wires for the power lead, but that's no guarantee. Then I noticed the nose headlight, which is a surface-mount LED, had a minus next to the yellow wire on its circuit board, **6**. So, red for positive!

The surface-mount LED in the nose had a resistor on its circuit board, so I left that alone. The other two LEDs didn't have resistors, so I added

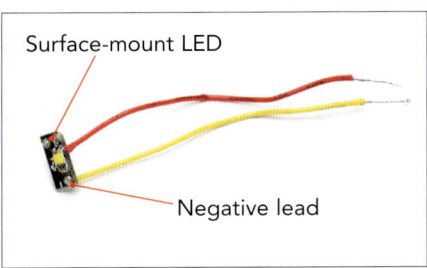

6. Eric needed to be sure which side of the LED was positive in order to wire the lights correctly. Fortunately, the circuit board for the surface-mount LED indicated the negative lead.

a 1KΩ resistor to each circuit on the common (+) side of the circuit, **2** and **4**. There were also a few spots where I had to splice wires so they'd reach the soldering pads on the decoder board.

I used heat-shrink tubing on all of the splices, and also on the connections between the resistors and the LED wires. Since I didn't want to risk overloading the function outputs for the decoder, I installed the cab and number box light on the function 5 circuit. It uses the same common connection as the nose headlight, which is where I attached the lead with the resistor.

Once I had all the wires ready to be soldered to the decoder, I went around and tinned them all. This makes the connection at the decoder quick, and cuts down on the chance of causing problems, such as flooding the board with solder and causing a short.

Another technique that helps is using the correct size of solder. I used 1mm rosin-core solder. Something even smaller would be fine, but the soldering pads on the Tsunami decoder aren't too small and are around the edges of the board, so it's not difficult to solder to them. Just be sure to have a clean, hot iron, and add just a touch of solder.

Before I soldered everything together, I placed a piece of double-sided tape on the bottom of the decoder to line up with the rear mounting pad. I put the tape over the hole in the board where it's meant to snap onto a locomotive frame in a drop-in installation. I left the backing on the tape so I could move the board around while I soldered the wires to it.

Testing and programming

With everything soldered together, I took the locomotive to the test track in our shop and fired up our NCE PowerCab. I used programming track mode to check the decoder and was able to read its manufacturer and model number, both good signs. I changed the address style from short to long and entered the locomotive's cab number for the long address.

Once I switched the throttle to programming on the main, I heard the sound of the GE FDL-16 prime mover coming through the speaker. I turned the headlight in the nose on, then observed the rear light switched on when I changed the locomotive's direction. Pressing F5 turned on the light on the front of the cab. Satisfied

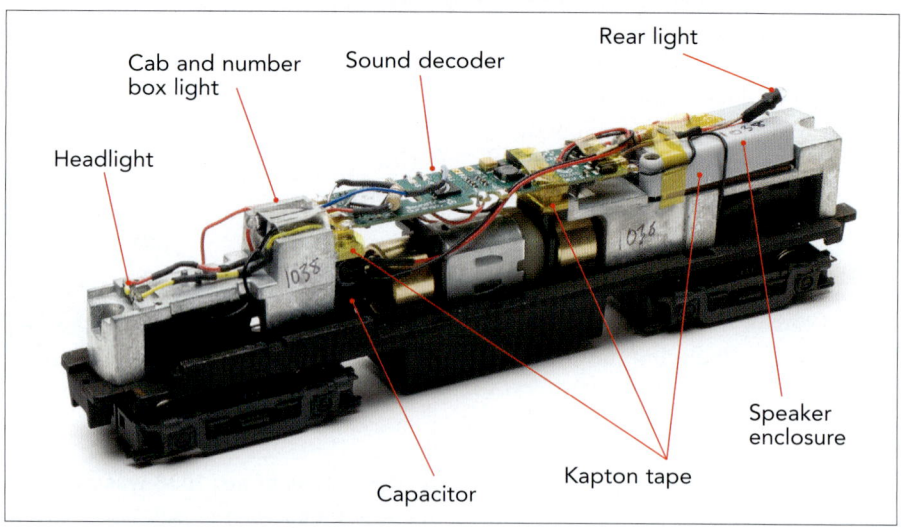

7. With the SoundTraxx Tsunami decoder installed in the Atlas U30B, all that was left to do was a little programming. Kapton tape keeps the wiring in place, and insulates the decoder from the frame.

that everything was working, I went back to the bench to put everything back together.

I checked to be sure the shell still fit together on the chassis with the decoder and wiring in place, pushing wires around to get them in the right spots. I then removed the backing from the tape and secured the decoder to the chassis. Next, I peeled the second piece of backing from the speaker gasket and secured the enclosure to the frame with the original screws.

The last thing to do was find a home for the capacitor. I hoped it would fit on top of the speaker enclosure, but then I saw empty space above the front driveshaft, behind the cab weight. There was room there, so I added a piece of double-sided tape and stuck it under the weight. I used a few more pieces of Kapton tape to tame a few wild wires, then slipped the chassis into the body shell. The finished installation is shown in **7**.

The final step was to tweak the decoder's configuration variables (CVs) to improve performance. I took the locomotive to the Eagle Mountain layout, since that's where it would be doing its work. Using the PROGRAMMING ON THE MAIN option on the layout's PowerCab, I started by increasing the value of CV3, Acceleration Rate, from the factory setting of 0 to 5, which didn't make a noticeable difference.

Programming on the main made it easy to change a CV, then immediately see the results. I kept adding to the value in 10-point increments until I had a level of momentum I liked. A value of 55 allowed the locomotive to start and transition between speed steps smoothly without adding too much delay. The Eagle Mountain is a small layout, so I didn't want a slow-to-respond locomotive running off the edge of the layout.

I followed a similar procedure for CV4, Braking Rate. Here, I ended up with a value of 33. I prefer to have the locomotive respond more quickly to braking requests for more control.

Installation of the decoder took a little more effort than with some other models, but by looking at the situation, it was possible to come up with a straightforward solution that used available materials and worked well. As with many aspects of model railroading, things can often be adapted to uses for which they weren't intended. Problem-solving can be a fun part of this hobby!

Adding sound to an N scale steam locomotive

1. Larry Puckett's friend wanted sound added to his N scale Bachmann Baltimore & Ohio EM-1 2-8-8-4. Larry took on the project, twice.

Installing sound decoders in N scale locomotives can be challenging unless the model was designed with this in mind. In some cases even when there appears to be plenty of room, there may not be room for extras like a stay-alive capacitor or a large speaker.

I ran into this problem installing a sound decoder in the Bachmann N scale Baltimore & Ohio 2-8-8-4 I used for this project, **1**. I'm going to show you two ways to do this installation, one that's relatively easy but doesn't produce the loudest sound, and a second approach that gives more sound by using a speaker in an enclosure.

My friend Bill Dye is a B&O fan who has several of the popular EM-1 2-8-8-4 models. These locomotives come with large tenders. When I offered to install a sound decoder, I figured it would be an easy drop-in project. After scanning the various decoder sound packages available, I ran across one for the EM-1 in the Digitrax sound depot.

I chose the Digitrax SDXN136PS sound decoder and PX108-6 Power Xtender stay-alive capacitor, **2**, and downloaded the EM-1 sound package, which is available from the Sound Depot on the Digitrax website (www.digitrax.com/sound-depot/list).

Preparation

Removing the tender body is easy. I backed out four screws and lifted off the shell, revealing a circuit board that runs the length of the inside and carries the motor-only decoder. A bundle of six wires from the engine terminate with a socket that mates with a six-pin plug on the circuit board.

Let me warn you about this socket. Printed on one side of this socket is the word "Up." However, as I later found after wasting a lot of time trying to figure out why the locomotive ran backwards and the headlight wouldn't work, "up" was actually "down."

At any rate, make sure to note which way the socket and the word "Up" are facing before disconnecting. Matter of fact, it's always a good idea to take a photo of a factory installation before doing any work on it.

With the wires disconnected, I was able to separate the engine from the tender. I removed the two small screws that held the circuit board to the tender

2. The Digitrax SDXN136PS sound decoder and PX108-6 Power Xtender stay-alive capacitor looked like a good fit. Try as he might, though, Larry couldn't fit the Power Xtender into the large tender. Larry also found the speaker too large, and the 8-pin plug wasn't used in this installation.

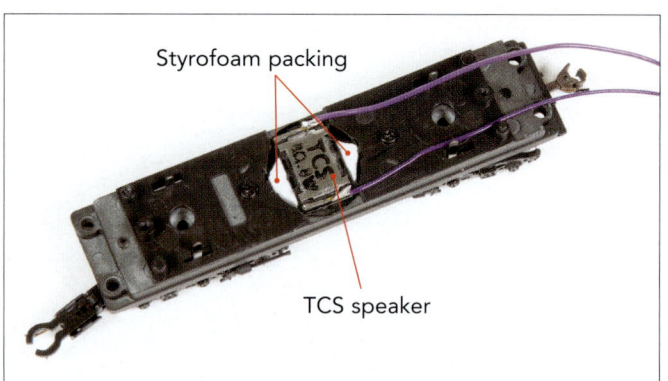

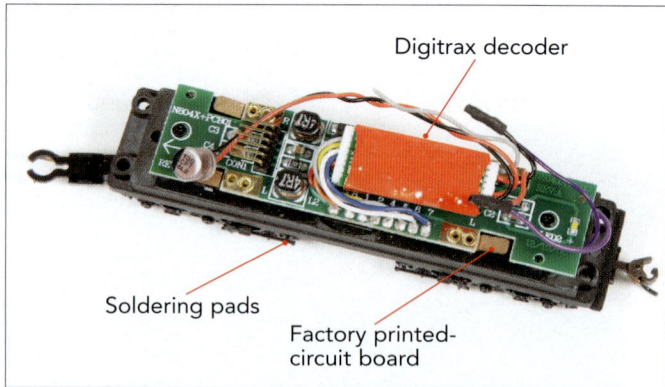

3. The Digitrax speaker was too large to fit the hole in the tender floor. To seal the sides of the rectangular Train Control Systems UNIV-SH6-C speaker in the round opening, Larry inserted a couple pieces of Styrofoam cut to fit.

4. Larry carefully removed the factory wires and soldered the wires from the Digitrax decoder in their place along the bottom of the circuit board. The speaker is mounted below the factory circuit board.

5. Larry soldered the wires from the decoder to the pins on the remaining part of the circuit board. The wire functions are (top to bottom) right rail pickup, motor +, headlight +, headlight -, motor -, and left rail pickup.

6. Larry connected a 2KΩ resistor to a surface-mount LED and attached them to the speaker enclosure beneath the rear light tube. The TCS sugar cube speaker and enclosure fit neatly in the rear of the tender.

floor and set them aside. The floor consists of a metal chassis overlaid with a plastic sheet to prevent shorts. A small circular opening for a speaker was cast into the floor.

Since the Digitrax oval speaker was too big, I chose one of the small sugar cube speakers I've been using the last couple of years for many of my HO diesel installations. I did some test fits and found a Train Control Systems (TCS) UNIV-SH6-C speaker, which is rated at 8 ohms and .8 watts, to be a good fit. I had to remove a little of the plastic liner to give me a drop-in fit.

A quick check with the circuit board confirmed there was just enough clearance. To seal the sides of the rectangular speaker in the round opening, I inserted a couple pieces of Styrofoam cut to fit, **3.** With the speaker firmly in place, I added a couple wires to the contacts and moved on.

The wires from the factory-installed motor-only decoder were soldered through holes in the circuit board. Fortunately, Bachmann included a list of the wire functions along with hole numbers, and the holes in the board are also numbered.

I carefully unsoldered the factory-installed wires and replaced them with the wires from the Digitrax decoder, **4.** I reinstalled the circuit board and connected the wires from the speaker to those on the decoder.

At this point I turned my attention to installing the Digitrax Power Xtender stay-alive. First, I tried placing it alongside the decoder, but it was too tall for the tender body. I also tried squeezing it under the coal load casting, but the connector from the engine was in the way there.

The rear of the tender looked like a possible option until I noticed there's a small surface-mount LED there that illuminates the rear light. So even in such a large tender, the space just wasn't adequate for both the Power Xtender and decoder.

With everything in place, I reconnected the plug from the engine and screwed the tender body back in place. On the programming track, I installed the sound package downloaded from the Digitrax Sound Depot. I used the Digitrax Soundloader software and the USB interface in my DCS240 command station (a Digitrax PR3 or PR4 USB interface and decoder programmer will also work).

After a little tweaking, I got the sound output maximized without creating any of the telltale crackling that's symptomatic of overdriving the speaker. On a small layout or in a quiet room the sound was reasonable, but when Bill ran the locomotive on his large layout with a room full of visitors, the sound almost disappeared, so I volunteered to make some changes. On to method two.

Second installation

After opening the tender again, I removed the circuit board and unsoldered the decoder wires from the first attempt. Next, using a razor saw, I cut the board just behind the six-prong connector. Using my volt-ohmmeter on the resistance setting, I isolated the right and left power pickup wires coming from the engine (these were the two outside wires).

The next step is a bit trickier, since I didn't have a connection to the motor or headlight. To keep from blowing out the headlight LED, I used a single AA battery, giving me 1.5VDC.

With the battery in a battery holder, I touched the wires to the pairs of socket openings, working from the outside in. The 1.5VDC was enough to turn the motor when I touched its contacts (the second and fifth wires). By observing the direction of the driver rotation, I could tell which wire was positive and which was negative. This meant the two inside wires were the headlight wires.

Since 1.5VDC isn't enough to illuminate a white LED, I switched to a pair of AA batteries in a holder, giving me 3VDC. Just to confirm this wouldn't blow the headlight LED, I first touched the wires to the backup LED still attached to the severed section of the circuit board.

Once I confirmed that was safe, I went ahead and touched the wires from the batteries to the center wires in the socket. When the headlight LED lit up, I noted the positive and negative wires. I added a 2KΩ resistor on the negative (white) wire to protect the LED and soldered the correct wires from the decoder to the pins on the remaining part of the circuit board, **5**. The resistor is necessary since the original one was on the section of circuit board I removed.

For the backup light I unsoldered the LED from the circuit board, added a 2KΩ resistor to the yellow (negative) wire, and glued the LED to the rear of the speaker enclosure next to the clear plastic light lens, **6**.

With most of the circuit board removed, I had to provide connections for power pickup in the tender. The trucks have metal prongs that stick up through the chassis floor and contact bronze wipers attached to the circuit board. The remaining section of the circuit board had the wipers for the front truck. I ran a wire between them to the prongs coming up from the rear truck. I then attached a wire from the decoder track wires (red and black) to the wipers on the circuit board, **7**.

I figured with most of the old circuit board removed, I now would have room for the Power Xtender. However, a quick test-fit with the tender body shot that idea down. The final arrangement did leave me with room at the rear end of the tender for the sugar-cube speaker and its intended enclosure, **6** and **7**.

One by-product of this new arrangement was the hole in the tender floor was now mostly covered by the decoder. To allow more sound out, I drilled a series of small holes in the plastic coal load, **8**. If the holes are distracting, they can be disguised by applying a thin coat of contact cement and then sifting on a fine coal load, making sure not to seal the holes. This arrangement seems to allow more sound to escape upward, whereas the hole in the tender floor directed the sound down onto the track and ballast.

After installing the tender body, I gave the locomotive a test drive, and the increase in volume was satisfying. You can't expect the same output as from a 1" high-bass speaker, but the output from these sugar cube speakers is always surprising.

To wrap this up, let me warn you that neither of these methods is easy. You'll need good soldering skills and a fine-tipped soldering iron. I also ended up setting my drugstore magnifying glasses aside in favor of an Optivisor with the highest magnifying lenses. However, with patience and the right tools, you should be able to do the same or a similar installation. I also strongly recommend testing this and any new installation on a current-limited programming track just in case of shorts. Good luck!

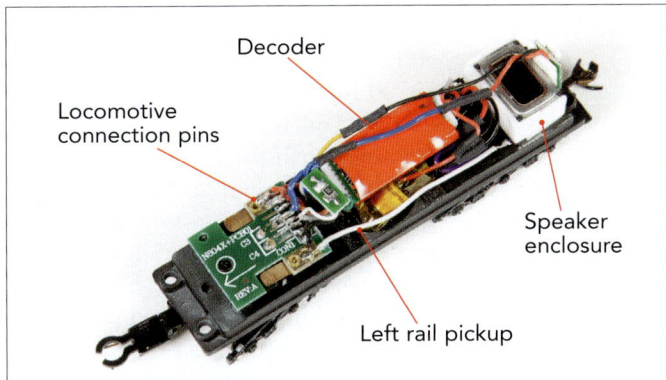

7. Larry soldered wires between the rear truck contacts and the bronze wipers on the remaining circuit board, then attached the wires from the decoder to the wipers. The white wire shown is for the left rail pickup.

8. To allow more sound out, Larry drilled a series of small holes in the plastic coal load.

3 Replacing a factory diesel decoder

1. Southern no. 2147, a Fairbanks-Morse H16-44, has its white class lights on as it couples to a string of cars in preparation for a shift as the Charlottesville local, an "extra," on Larry Puckett's HO scale Piedmont Southern layout.

I've had several requests from readers about installing a sound decoder in first-generation Atlas Fairbanks-Morse H15-44 and H16-44 models, **1**. These were originally released around 2003 and were available with Lenz mobile decoders. The interesting feature of these Lenz decoders was the ability to turn the classification lights on and off. When turned on using Function 6, they displayed a green light on the forward end with a red light on the rear. These colors reversed automatically when the locomotive changed direction of travel.

As since-retired MR senior editor Jim Hediger pointed out in his August 2003 product review, this choice of class light colors isn't realistic in prototype operation. Locomotives of that era ran with the class lights off if leading a scheduled train, white class lights and/or flags if leading an extra, and green lights and/or flags to indicate a following section of a scheduled train. Red was used as an end-of-train marker, with a red light, flag, or reflector disc displayed at the rear end of a train.

On recent releases of this model, Atlas used an Electronic Solutions Ulm (ESU) LokSound 21-pin sound decoder and retained the red-green class lights. To more closely follow prototype practices and keep this project simple, I installed independently controlled white/green class lights and a LokSound decoder in my older locomotive.

Because I used the same sound project as those in the new factory-installed decoders, my locomotive should be compatible with the current release. I used white light-emitting diodes (LEDs) since I only operate regularly scheduled trains and extras, but if you plan to operate some trains in multiple sections, you'll want green LEDs.

In this installation I used a no. 73700 LokSound Select Direct decoder, which is a different configuration than the 21-pin decoder Atlas uses. However, the sound project I used is the same. Most dealers sell these universal LokSound decoders with generic diesel sounds. So unless you plan to install the sound package yourself, when you order the decoder make sure the dealer installs the correct sound project for you. Most dealers will do this, but you must ask. In this case I chose the 93449 Atlas H15-44/H16-

44 sound package from the LokSound North American & Australian Factory Equipped Sound files on the ESU website.

Installation

I slipped the shell off the chassis and surveyed the situation. I left the class light LEDs attached to the board, since they were soldered on. I did, however, disconnect the wires to the motor, the forward and rear light bulbs, and the power pickups. With all the wires disconnected, I lifted out the decoder by using a small screwdriver to disengage the plastic clips that secure the decoder.

I checked the orientation of the new decoder board so the front end faced the front end of the locomotive. It's important to do this, since some railroads designated the long-hood end of these locomotives as forward. I connected the motor leads to the terminals on the board and slid it into place on the plastic mounts. Using the plastic caps on the wire leads makes it easier to reverse the wires if you later find the locomotive doesn't go in the correct direction. I also attached the pickup wires from the trucks at this point.

I installed white 3mm LEDs for the class lights and headlights. The LEDs are available in white or green, but to keep things simple, I ordered sunny white versions from Richmond Controls (www.richmondcontrols.com). I chose the sunny white, which have no blue cast at all. Resistors are unnecessary, since the LokSound decoder board comes with 2.2KΩ resistors on the function outputs.

Installing the 3mm LED replacement headlights is simple, since the chassis has a slot designed for them. I soldered red and black wires to the positive and negative legs of the LEDs. The longer leg is positive, plus there's usually a flat spot cast into the negative side of the plastic lens.

I placed heat-shrink tubing over the solder joint and all exposed metal, as the LEDs will be in direct contact with the chassis. Finally, I added a longer piece of 1/8" heat-shrink tubing

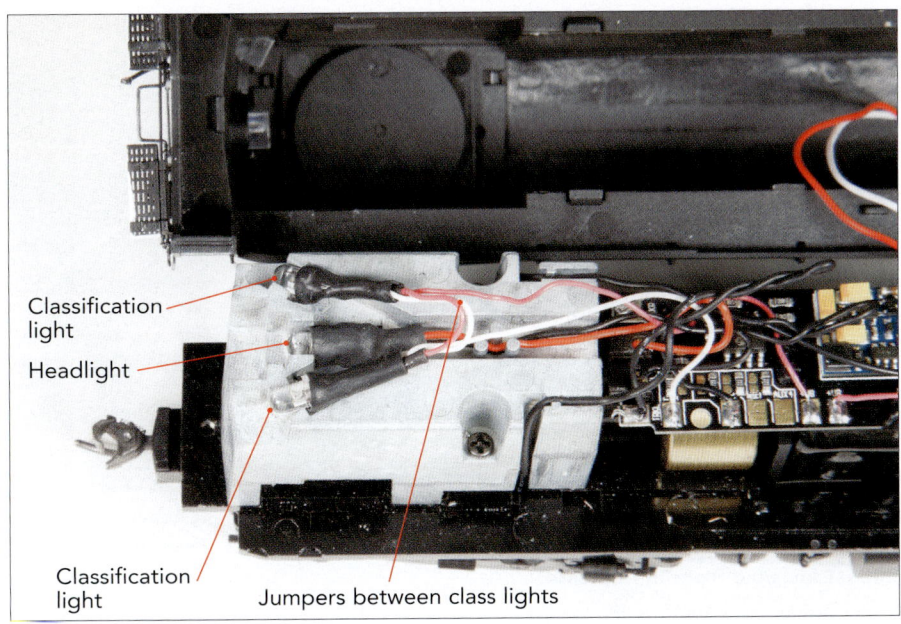

2. First Larry installed the LED headlight. To simplify the class light wiring, he added jumpers between the LEDs at each end, then ran a single positive and negative wire from the pair of LEDs to the function connections on the board.

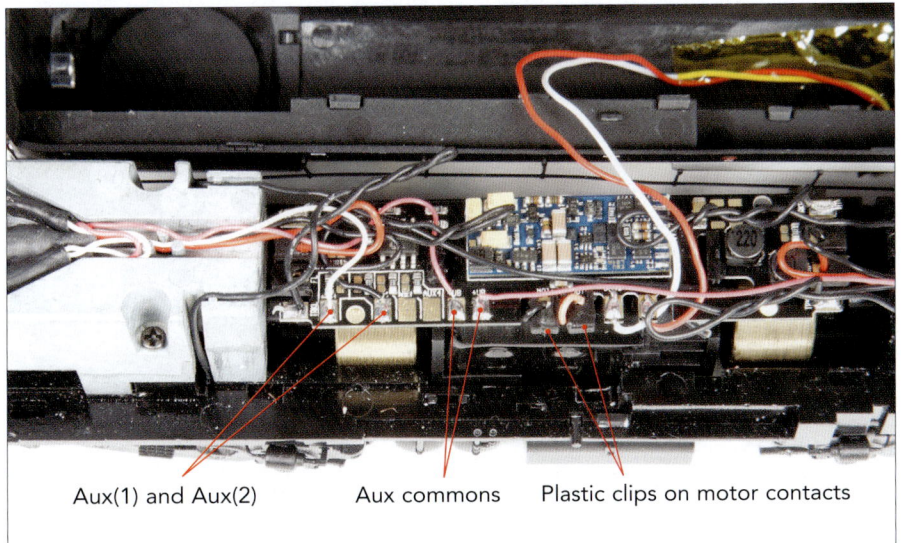

3. Larry connected the class lights to the function contacts labeled Aux(1) and Aux(2). They're controlled by Function 6 on the throttle.

over the base of the LED and down over the wires. I installed the LEDs into the slots in the chassis in the same orientation as the ones I removed, **2**. Next I connected the wires to the decoder terminals at each end of the board, observing proper polarity.

The classification lights required a bit more work. For each end of the model I soldered a jumper wire between the positive LED legs and added one long wire creating a single common, then placed heat-shrink tubing over the solder joint all the way up to the base of the LED, **2**. I did the same for the negative leads, giving me only two wires for each LED pair. I added larger diameter tubing over the LED and wires just like with the headlights.

The LokSound decoder has four auxiliary function contacts on one side of the board, **3**. These are designated Aux(1), Aux(2), Aux(3), and Aux(4) for programming the decoders. I used Aux(1) and Aux(2) for the white class

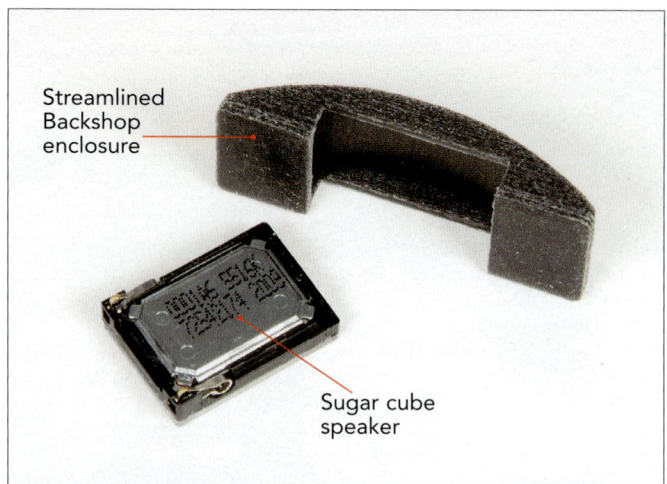

4. Streamlined Backshop's sugar cube speaker with curved enclosure fits the inside of a locomotive cab roof, making it almost invisible from the outside.

5. Larry installed the speaker in the cab using a small rectangle of double-sided foam tape and soldered the connections to the board.

LEDs on the front and rear ends, respectively. I connected the negative wires from each LED pair to the appropriate Aux function pad. I then connected the positive wires to the function common pad—there are two of these next to the Aux function pads, and it doesn't matter which you use.

With all the other connections completed, I turned to installing a speaker. As I showed in the April 2018 *Model Railroader*, Streamlined Backshop offers sugar cube speakers with enclosures curved to fit the inside of a locomotive cab roof, **4.** After gluing the speaker into the enclosure using cyanoacrylate adhesive (CA), I soldered wires to the contact prongs. Using a small rectangle of double-sided foam tape I attached the enclosure to the inside of the cab roof. With the speaker installed, it was a simple matter to attach the speaker wires to the contacts on the decoder board, **5.**

With everything installed I was ready for the programming track. I changed the address to match the locomotive number, and other CV settings to match my other locomotives. The programming for the class lights is already in the decoder. Because the program for the Atlas models uses Aux(1) and Aux(2) to turn on the class lights on opposite ends of the locomotive, I was able to use that without making any programming changes.

With the LEDs wired as above, the class lights come on when F6 is pressed, and automatically reverse when the locomotive changes direction. I told you this would be simple!

Adding sound to an older HO Atlas Alco RS-1

1. Southern RS-1 no. 405 heads back to Charlottesville with a local on Larry Puckett's HO scale Piedmont Southern layout. Larry added a Digitrax sound decoder with a "sugar cube" speaker to the model.

Many years ago, I bought an Atlas HO scale Alco RS-1, one of the first batch produced by Kato with the smooth-as-silk mechanism, **1**. I recently painted it, then looked for an appropriate decoder for it. I chose a Digitrax 16-bit sound decoder, the SDXH166D. It's rated at 1 amp continuous and 2 amps peak motor current, and has up to six light functions.

With 16 megabits of onboard sound storage, the decoder works with Digitrax's proprietary 8-, 12-, and 16-bit sound projects, can play four sounds simultaneously, and provides an 8Ω, 1W speaker output. Because of the configuration of this model, I used a compact "sugar cube" speaker in place of the part supplied by Digitrax.

Each decoder comes with a factory-installed speaker and sound project that includes eight steam and diesel locomotives. If you have a PR3 programmer, you can download and install other sound projects from the Digitrax website (www.digitrax.com). Coincidentally, the decoder comes with an RS-1 as the factory-installed sound project, so it was a perfect choice for me.

There are two options for installing these decoders: attach the decoder to the inside of the shell, or attach it to the motor. No matter which one you use, you're going to have to install a connector or two for the speaker and lights, or the motor and track pickup wires. I went with the second method after some test fits using the first method didn't work out.

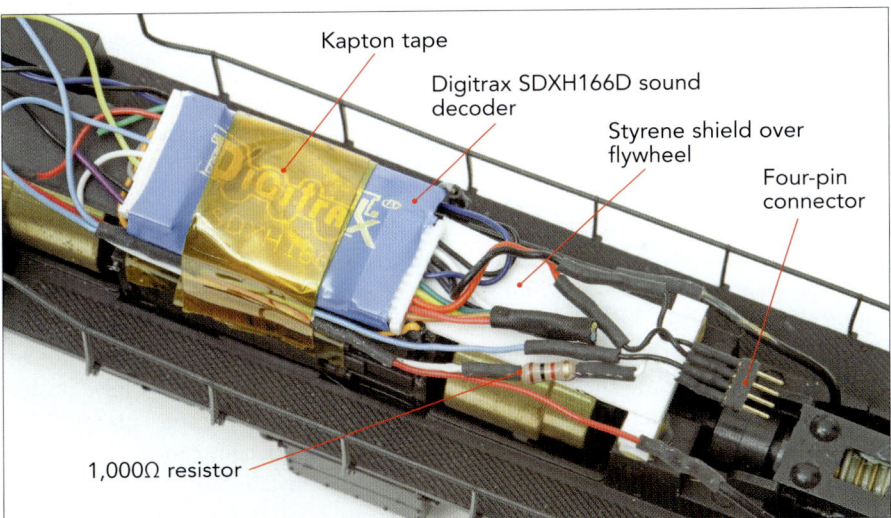

2. The decoder is mounted on top of the motor using Kapton tape. Two- and four-pin wiring harnesses connect the decoder to the lights and speaker in the shell. The strip of styrene prevents the wires from rubbing against the flywheel.

Installation

To begin, I removed the shell and pulled out the light board and light tubes. I applied a couple layers of Kapton tape to the top of the motor for added protection against short circuits and taped the decoder to the motor using another strip of Kapton tape, **2**.

I soldered the gray and orange wires

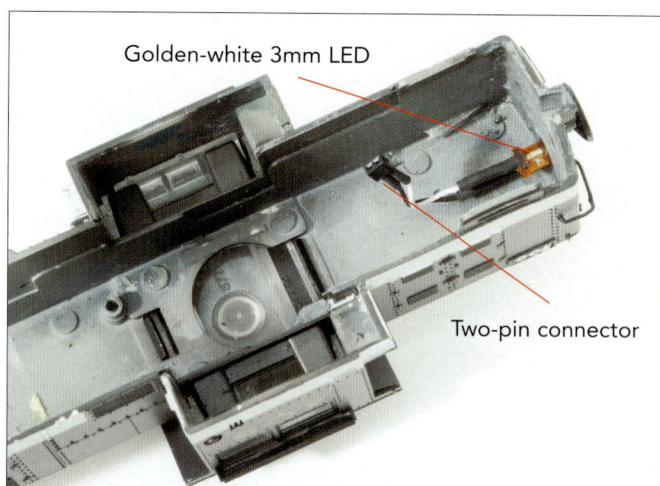

3. The LEDs are glued to the inside of the shell directly behind the headlight castings. The two-pin connector on the short hood end makes final hookup quick and easy. A four-pin connector is at the other end.

4. Although these "sugar cube" speakers (three from Streamlined Backshop and one from Tony's Train Exchange) are extremely small, ranging in size from 8 x 12mm to 13 x 18mm, they put out significant sound volume in frequencies useful for model locomotives.

from the decoder to the left and right motor leads, respectively. Then I spliced the black decoder wire to the left track pickup wires and the red wire to the right side wires. I installed sheets of styrene over the flywheels to prevent wires from rubbing against them. At this point I gave the model a test run to make sure the locomotive ran in the correct direction.

I replaced the original headlights with two-light Pyle castings, requiring a different light source front and rear. I like the color of golden-white 3mm light-emitting diodes (LEDs) from Richmond Controls, and opted to use them. I soldered wires to the positive and negative leads using different colored wires so I could keep track of the negative one, which also got a 1KΩ resistor (the short leg is negative). Remember to always protect all solder joints using heat-shrink tubing.

To mount the LEDs, I sanded each rounded end flat, then, using cyanoacrylate adhesive (CA), attached the LEDs to the inside of the shell behind each headlight, **3**. The lenses are Testor's acrylic Clear Parts Cement with a touch of Glosscote on the outside to protect them.

New locomotive models designed with a sound decoder installation in mind usually have room provided for a speaker. However, my locomotive isn't new, and the speaker Digitrax included wouldn't fit. So I used a "sugar cube" speaker, **4**, available from (among others) Streamlined Backshop (www.SBS4DCC.com) and Tony's Train Exchange (www.tonystrains.com).

These speakers originally were made for use in cell phones, tablet computers, and similar devices, so they're very small. They range in size from 8 x 12mm to 13 x 18mm and use plastic enclosures. The folks at Tony's Train Exchange have been able to measure usable output frequencies as low as 200 Hz, so you can expect good low-frequency sound, and they're capable of putting out about 72 Db (I have posted videos with sound samples on my website, www.dccguy.com).

For this installation, I test fit both 11 x 15mm Streamlined Backshop and 13 x 18mm Tony's Train Exchange speakers with about equal sound quality and volume. I finally went with the smaller one to simplify installation.

Most sound decoders are rated for 1W at 8Ω. However most of these sugar cube speakers have a nominal rating of 0.5W to 0.7W. Although they also carry a 1W maximum rating, you have to be careful powering them, as this maximum rating is based on tests for one second on, followed by one minute off. So I recommend keeping your output near the nominal rating.

To do this with a 0.7 watt speaker and 1 watt decoder, set the master volume at 70 percent of output. Then adjust the individual sound levels, knowing that no matter how high you set them, they are unlikely to blow the speaker.

The Streamlined Backshop speakers have to be glued to their enclosure (I used CA) whereas the Tony's Train Exchange speakers are simply press-fit into their enclosure. It's important to get an airtight fit to provide the best sound quality.

I soldered wires to each of the small metal spring clip contacts on the back of the speaker, being careful not to create any solder bridges to the metal back plate. As long as you're only installing one speaker, polarity doesn't matter. To prevent the wires from breaking loose, I wrapped a thin strip of electrical tape around the side of the speaker and wires.

Next, I cut a small rectangle of double-sided foam tape, attached it to the back of the speaker enclosure, and eased it into place behind the forward headlight LED, **5**.

With the speaker and both LEDs installed in the shell, I now needed to provide for the wiring connections to the decoder. While you can solder the speaker and LED wires directly to the decoder wires, I prefer to use multi-

5. A thin strip of electrical tape around the side of the speaker and wires helps to stabilize them. Double-sided foam tape seats the speaker in the front of the unit, with a four-pin wiring harness and connector for the speaker and LED wires.

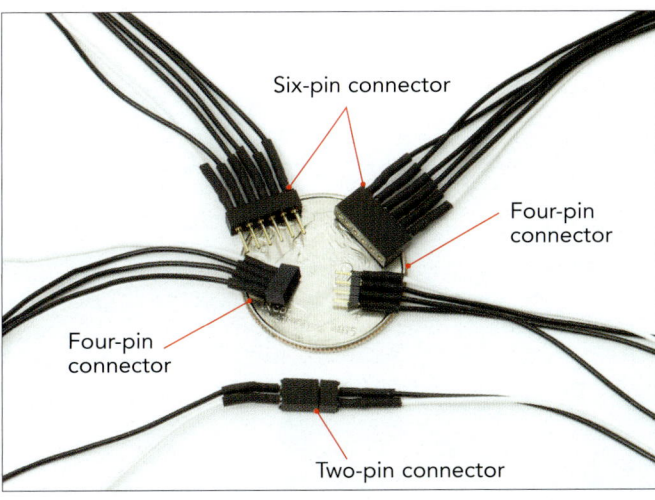

6. Wiring harnesses like these from Train Control Systems have flexible wires and small connectors that come in two-, four-, and six-pin configurations. Miniatronics and others make similar harnesses.

pin connectors. This makes it easy to separate the shell from the chassis if maintenance is required.

I used mini-connectors made by Train Control Systems, which are small, have flexible wires, and one white wire, **6**. The white wire makes it easy to maintain proper polarity when connecting components like LEDs. I soldered a four-pin mini-connector to the four wires from the forward LED and speaker, making sure to attach the white wire to the negative LED lead, **5**.

I used a two-pin connector for the rear LED, **3**, but in hindsight, one six-pin connector would've simplified things. Finally, I hooked the connectors up to their mates and eased the shell into place.

Programming the decoder is straightforward. There are no indexed configuration variables (CVs) to deal with, and I was able to program it on the service mode track without a programming track booster. Digitrax offers an extensive list of light effect options. I set the headlights up so both stay on regardless of direction of travel and activated Rule 17 dimming, which automatically dims the headlight when running in reverse.

Changing some sounds like the air horn is easy. The default horn for the RS-1 is a single chime "blat" type, whereas my locomotive has a multi-chime Nathan M3. After searching through the manuals, I couldn't find a list of the available horns, so I tested all of them to find one that sounded close to my M3.

One disappointment was that sound functions can't be remapped to different buttons without rebuilding the sound projects themselves. Although this can be done using the free Digitrax SoundLoader software, it may be more than many users will want to attempt. But if you decide to install these decoders in a number of locomotives, it might be worth learning.

One conflict created by this limitation is that both switching speed and prime mover notching are controlled by function 6. Consequently, you can only have one activated at a time, otherwise every time you select switching speed the locomotive will notch up, and vice versa.

After 20 years of making decoders, Digitrax has mastered the art of designing speed control algorithms. The Digitrax LocoMotion system includes back-electromotive-force control (back-EMF), the preset speed tables, switching speed, and various other built-in features. Turning on switching speed drops top speed by about 50 percent and momentum by 25 percent. This allows you to run at mainline speeds, then drop to lower speeds with precise control when switching. Three throttle modes are provided: automatic with eight regularly spaced notch settings, semi-automatic with the ability to manually advance notch settings above and below the throttle setting, and full manual, where notches are independent of throttle settings. Overall, locomotive response is as smooth and predictable as with any current decoder I've tested.

So how does it sound? I was impressed with the way the decoder brought my locomotive to life. The RS-1 recordings have a clear, crisp tone. The decoder offers the usual shopping list of user-controlled sounds along with a selection of air, brake, and other sounds that play automatically.

The only sound that could've been louder was the whine of the turbo. You can hear it if you listen closely, but in an operating session with other noises it was barely distinguishable. What I liked best was the familiar rattle of the Alco prime mover—I've always thought it sounded like someone shaking a couple ball bearings in a tin can. I definitely plan to use more of these decoders in my expanding locomotive fleet.

3 Replacing an old HO steam sound decoder

1. Large articulated locomotives like this Clinchfield Challenger 4-6-6-4 make an impressive presence, and their sounds overwhelm other locomotives. But this model from 2004 didn't have all the features available today.

One of the realities of DCC decoders is that as technology advances, manufacturers keep updating decoders and adding new features, functions, and sounds. Consequently, even an expensive model that was at the cutting edge when it was released may be out of date in a few years.

I ran into that problem with my Athearn HO Genesis Challenger 4-6-6-4 lettered for the Clinchfield, 1. These models were first released around 2004, with an early decoder that lacked a lot of features we've come to expect in sound-equipped locomotives today.

My model appears to be pretty much the same externally as the most recent releases, so replacing the whole model seemed a waste of money. Instead, I tackled the job of pulling out the old decoder and replacing it with a new one. Although my model is likely different from those you might have, the steps in replacing a decoder should be similar.

Right off I ran into the problem of getting to the decoder inside the tender. There are screws under the coal bunker and under hatches on the tender deck. The rear of the tender has to be removed after taking out two screws on either side of the coupler and one behind it. Finally I was able to slide the tender body off the chassis, revealing the decoder, 2. The decoder is shaped specifically to fit this tender, even serving as a base for the light-emitting diode (LED) that illuminates the locomotive's rear light. The decoder is held in with two screws.

Another problem you may run into is that factory-installed decoders may not follow the National Model Railroad Association wiring standard. Therefore, the first task is to figure out what each wire is connected to. Before jumping into a lengthy diagnosis of the wire connections, do an internet search to see if anyone has worked on the same locomotive model and posted the wiring information. I got lucky and found a detailed description of all the wire connections for this decoder, saving me some time.

However, it's still a good idea to verify these connections. I discovered some were reversed, including those for the power pickup and the LED headlight.

If you can't find the wiring information for your specific model, here are a few tips for tracing wires. The easiest to trace are the wires going to the right and left track power pickups: Just follow the wires up from the trucks to their solder points on the decoder and mark them with little strips of tape.

If yours are hidden, you can set a multimeter for resistance and check for continuity between each wire and the wheels on each side of the tender until you find the ones that match.

Next, look for the headlight wires. If it's a white LED, it's probably safe to assume it operates at about 3VDC. However you may want to start with 1.5VDC and use a single AA battery to test pairs of wires and see if the headlight comes on. If that doesn't work, try two AA batteries, giving you 3VDC to work with. If during the process the drivers start to turn, you've found the motor leads.

Some steam locomotives may also have a smoke generator, which can complicate the process. In some cases you may even need to open the boiler to trace the individual wires.

After identifying the wires, I cut them just above their solder joints, removed the screws holding the decoder in place, then lifted the decoder out. This left a small circuit board that connects to the electrical socket at the front of the tender.

New decoder

I chose a TCS WOWSound Version 4 decoder with a KeepAlive capacitor system. This decoder offers excellent sound and motor control, a large selection of whistles (61) and bells (47),

2. The decoder used in the first generation Athearn Genesis Challengers filled the large centipede tender.

Tender circuit board — DCC decoder — Back-up light

58

3. Both the TCS WOWSound decoder and Keep-Alive capacitor pack fit comfortably in place of the older decoder.

4. A styrene shelf holds the light-emitting diode (LED) at the right level to illuminate the backup light. A 1,000Ω resistor protects the LED.

and many other steam sounds.

In addition, the prototype throttle mode adds realism to the engineer's job. The standard format decoder and KeepAlive fit neatly on the narrow frame.

To prevent any chance of damage to the decoder, I soldered the wires from the decoder's harness to their respective wires on the remaining small circuit board before connecting the harness to the decoder. I prefer to do this because it's recently come to light that some soldering irons can leak current through their tips and damage sensitive parts on decoders during the soldering process (see page 40, and also see my website, www.dccguy.com). Once the connections were all made, I attached the decoder to the frame using double-sided foam tape, **3**.

Because the old decoder apparently had a resistor to drop the current for the LEDs, I needed to install replacements. I soldered a 1000Ω, ½W surface-mount resistor on the negative wire to the headlight LED. Since the back-up light LED was mounted on the old decoder, I had to come up with a replacement.

I cut a piece of styrene, made a small hole in it, and slid it down on the rear screw post. The taper of the post combined with the size of the hole allowed me to control the position of the styrene. After soldering wires to an Ngineering incandescent LED,

I attached it to this styrene platform using a small rectangle of double-sided foam tape. The platform also provided a convenient place for a 1,000Ω surface-mount resistor, **4**.

The final step was installing a new speaker. Two flat 32mm diameter speakers mounted in plastic baffles and suspended from the top of the tender comprised the factory installation. I knew that I wanted to replace them with a 28mm high-bass speaker like the one I used in a previous installation. The speaker is from Streamlined Backshop (www.sbs4dcc.com) and the enclosure from Micro-Mark, which also sells high-bass speakers in a variety of sizes.

I drilled a small hole in the side of the enclosure near the bottom and inserted the two purple speaker wires from the decoder. After soldering the wires to the solder pads on the speaker, I ran a bead of cyanoacrylate adhesive (CA) along the lip inside the enclosure and seated the speaker. Remember to test the speaker before gluing it in place!

I considered gluing the enclosure to the top of the tender, then realized I could use the old mounting screw to do the job. I test-fit the enclosure and made a mark on it from the outside, then drilled a 5/64" hole in the enclosure, being careful not to drill into the speaker.

Because the location of the speaker interfered with the mounting clips for the coal bunker insert, I added a couple strips of double-sided foam tape as spacers, creating just enough room for the clips. I then ran the screw into the back of the enclosure from the outside of the tender, and covered it again with the hatch casting.

After reassembling the tender, I plugged in the harness from the locomotive and placed it on the programming track. I created a new DecoderPro roster entry and changed the address, then moved the Challenger back to the main track and proceeded to program the decoder on the main.

After setting the locomotive for heavy steam with articulation, I ran through the available whistles—all 61 of them. Since there were no Clinchfield whistles or bells to choose from, and no one was able to suggest a similar selection, I assumed that a Union Pacific three-chime whistle would be a reasonable alternative. I also selected a Chesapeake & Ohio bell.

The improvement in the bass frequencies because of the high-bass speaker is amazing, especially with the heavy steam chuffs and a steamboat whistle like the Union Pacific prototype I selected. To hear the sounds of this updated Challenger, visit my website (which also includes a description of the process I went through to replace the orange headlight LED with a golden white one).

3 Choosing speakers for sound installations

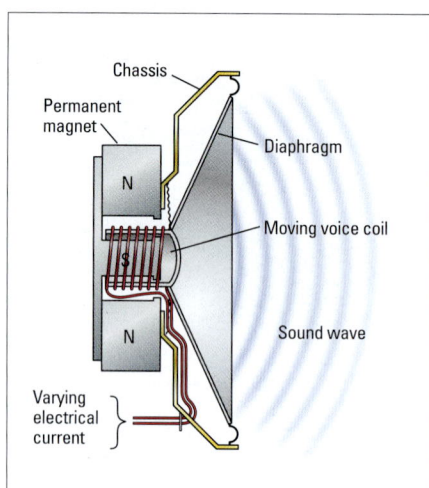

1. Speakers are complex little devices. This illustration shows the components used to produce sound.

Several factors need to be considered when choosing speakers for a sound installation. These factors range from the type of locomotive to the size and electrical characteristics of the speaker to the way you install and wire the speaker(s).

First, let's look at what goes on when sound is generated by a speaker. The electrical current carrying a sound signal varies in proportion to the sounds it carries. At the speaker, those varying electrical impulses pass through a small coil of wire (often referred to as the voice coil) attached to the back of a diaphragm made of plastic, paper, or other flexible material. The parts of a speaker are shown in **1**.

The electrical impulses create a varying magnetic field in the voice coil, which is repulsed by a permanent magnet on the back of the speaker. This repulsion forces the diaphragm outward, pushing a wave of air ahead of it—that's the sound wave that hits our eardrums.

For every forward movement of the diaphragm, there's a similar rearward movement as it returns to its original position. As the diaphragm moves backward it pushes air, creating another sound wave from the backside of the speaker. Since the two sound waves are created at slightly different times, they are mirror images, and said to be out of phase. When they mix on the way to our ears, they can cancel each other out. This is one reason an open speaker produces less sound than an enclosed speaker.

The important factor in all this is that if you can keep the two sound waves apart, you can prevent the deadening effect. What's also important is that since sound waves are generated by both sides of a speaker, in many cases it usually doesn't matter which side of the speaker is facing out.

By installing the speaker in an enclosure of sound-trapping material, one of the sound waves is contained, and can't interfere with the other one. Thus you get the full sound being generated.

Many manufacturers now offer speakers and compatible enclosures for just this purpose. You can also make your own enclosure out of old pill bottles, film canisters, styrene sheet, and other materials, **2**. A steam tender makes an excellent speaker enclosure, and some models come with a spot for a speaker with holes in the floor to let the sound out.

When it comes to enclosure design, the rule of thumb I've always tried to follow is to use an enclosure as deep as the diameter of the speaker. Of course, this may not always be possible, but I try to make it as deep as will fit.

Enclosure design can be complex, especially when it comes to ported or bass-reflex enclosures, **3**. These have a hole that allows some of the sound waves produced by the backside of the diaphragm to exit via a specially designed port. This design increases the bass frequencies, which is great for both diesel and steam locomotives.

Speaker size is another important factor in the quality and volume of the sound produced. Generally, bigger speakers produce more bass and more volume than smaller ones. There are also specialized designs, such as high bass, that can improve bass performance.

One of the most exciting developments are "sugar cube" speakers, designed for use in cell phones and tablet computers, **3**. These speakers produce much better sound than might normally be expected given their small size, as long as they have a proper enclosure.

Another important aspect of matching decoders and speakers is the ohm and wattage rating. Most decoders today are rated at 8Ω, 1 watt. For best performance, decoders should be matched with speakers of the same ratings. With respect to wattage, keep in mind if you operate a speaker at a higher wattage than it's designed for, you run the risk of burning out the voice coil. However, if two speakers are wired in either series or parallel, they will each receive half the wattage put out by the decoder.

So you can use two 0.5W speakers safely with a 1 watt decoder. If you do

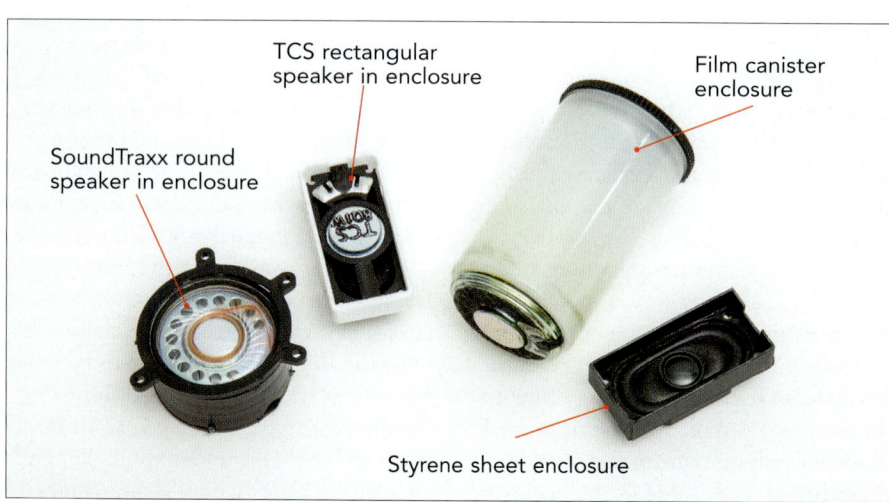

2. Although there are many commercial speaker enclosures, like the ones at left from SoundTraxx and Train Control Systems, simple speaker enclosures can be made from old film canisters or styrene sheet.

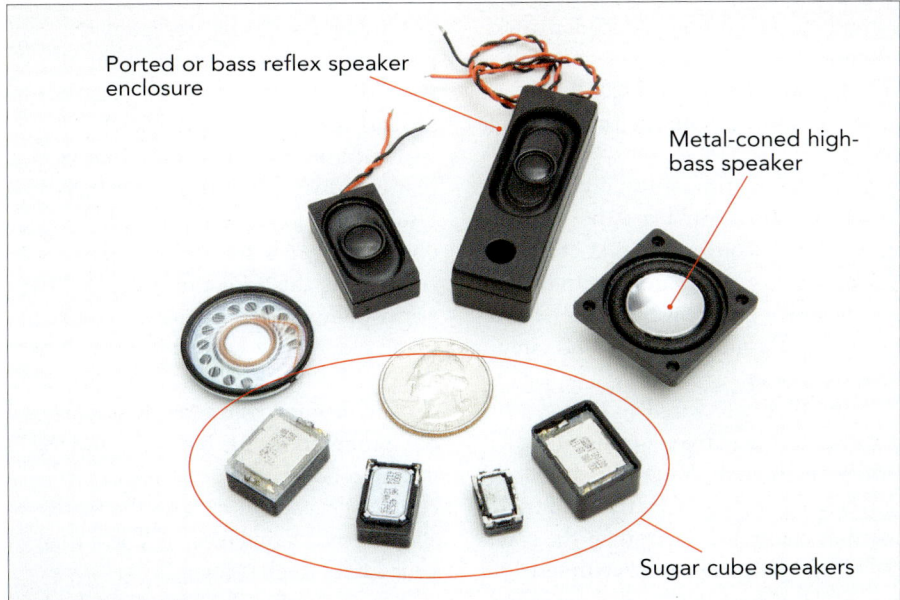

3. A variety of speaker types and sizes are available, with and without enclosures. Round and oval speakers are commonly used in commercial installations and are a favorite choice for do-it-yourselfers.

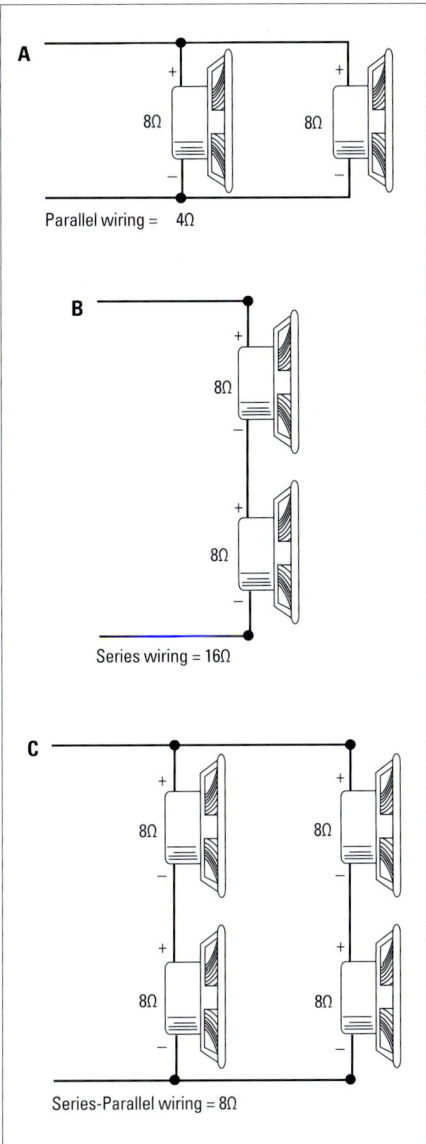

4. Multiple speakers can be connected to a decoder, wired either in (A) parallel, (B) series, or (C) a combination of the two. This allows matching the ohm rating of the speakers to that of the decoder. When connecting multiple speakers, it's important to maintain polarity. *Original drawing courtesy of SoundTraxx*

use a 0.5 or 0.7W speaker with a 1 watt decoder, you can turn down the master volume control to 50 percent or 70 percent, respectively, to limit the wattage.

Most speakers have both a continuous and a maximum wattage rating. For example, a speaker with a 0.5W continuous rating may have a maximum rating of 1W. However, that maximum rating may only be for a short time, with a longer off period between peaks. You can also use speakers with a wattage rating larger than the decoder's rating, such as a 2W speaker with a 1W decoder.

An 8Ω decoder can usually drive a speaker rated anywhere from 4Ω to 16Ω. It might seem counter intuitive, but it's actually better to match an 8Ω decoder to a 16Ω speaker rather than a 4Ω speaker to avoid overdriving the amplifier.

You can use this to your advantage when installing multiple speakers once you know that ohms add when the speakers are wired in series and are halved when wired in parallel.

For example, two 8Ω speakers in series would be rated at 16Ω, and in parallel would be 4Ω, **4**. By using two 4Ω speakers in series, you would be back to 8Ω. You can also wire four 8Ω speakers in a combined series and parallel pattern that results in the decoder "seeing" only 8Ω. Streamlined Backshop (www.sbs4dcc.com) has an excellent tutorial on this on its website.

With multiple speakers, it's important to ensure polarity is maintained, **4**. If not, one speaker will produce sound from the front of the diaphragm while the other produces it from the rear, resulting in poor quality. Some speaker wiring contacts are marked for polarity.

If not, then it's usually safe to assume speakers of the same size, rating, and manufacturer have the same polarity. For example, the left contact may be positive on all speakers of the same type and size produced on a given production line. However, it isn't usually safe to make this assumption for dissimilar speakers or ones from different manufacturers.

Something to look out for when setting sound levels is a type of distortion called clipping. Clipping can occur when you attempt to drive the amplifier to a power output greater than it's designed for by using speakers without enough impedance. The result can be sound distortion. Often a crackling sound is the only warning you'll get before everything goes silent.

The result can be overheating and failure of the amplifier or speaker. I once blew a SoundTraxx decoder by overdriving it. Fortunately, SoundTraxx was able to replace the amplifier chip. I learned my lesson from that early experience—it's best to match decoder and speaker output ratings.

3 Stay-alive modules

Stay-alive devices have been a part of DCC since about 2005, when Lenz released the first ones. However, these devices didn't really begin to get much attention until Train Control Systems (TCS) introduced its version in 2013. Since then, most manufacturers have offered similar devices.

"Stay alive" and "keep alive" are commonly interchanged terms. Train Control Systems trademarked its stay-alive device the "Keep Alive" first, and everyone else had to come up with different names like CurrentKeeper, Power Xtender, No Halt, PowerPack, and the newest addition, KeepRolling, **1** and **2**.

What they do; how they work

A stay-alive is a small electronic device designed to help keep your locomotive rolling over electrically dead switch frogs and dirty sections of track. They have become more important as sound decoders have grown in popularity, because the illusion of reality suffers if the locomotive's sound keeps cutting out.

To understand how a stay-alive works, you first need to understand something about how a decoder works. Decoders operate on direct-current (DC) power. However, DCC track power is a form of alternating current (AC).

A bridge rectifier—usually made up of four diodes in a specific orientation—converts the AC to DC. Track power goes in and positive and negative DC power comes out, powering the mobile and sound circuits of the decoder.

Stay-alives are wired to the positive and negative outputs of the bridge rectifier, **3**. When the decoder is receiving track power, the bridge rectifier charges capacitors in the stay-alive. Then, if power is interrupted, power flows back out of the stay-alive capacitors and powers the motor control, lighting, and sound circuits.

Most modern decoders have sockets where the stay-alive can be plugged in, making installation easy. However, on decoders that were manufactured before stay-alives became common (2013 or so), you have to make the connection yourself. This means finding the positive and negative contacts on the bridge rectifier diodes.

The positive connection is easy; just use the blue wire that serves as the positive common connection for all your functions. Finding the negative or ground contact is a little trickier.

First, you must locate the four diodes that comprise the bridge rectifier, **3**. Usually the negative ground will be one of the two sides of the diodes comprising the bridge rectifier. In **3** the arrow points to the negative contact on an older Lenz LE080 decoder. You can test for this location on a powered decoder using a voltmeter by placing one probe tip on the blue wire and the other probe tip on one of the diode contacts. If you get it right, the meter should read as much as 1.5V less than your DCC track voltage.

Once you find this contact, the blue wire from the stay-alive should be soldered to the blue common wire on the decoder and the other wire (usually black/white) to the negative contact point. The National Model Railroad Association (NMRA) has designated blue as the positive wire and black/white as the negative wire. However, these colors may vary on some devices.

One of the best resources for determining these contacts on other decoders is Marcus Amman's website

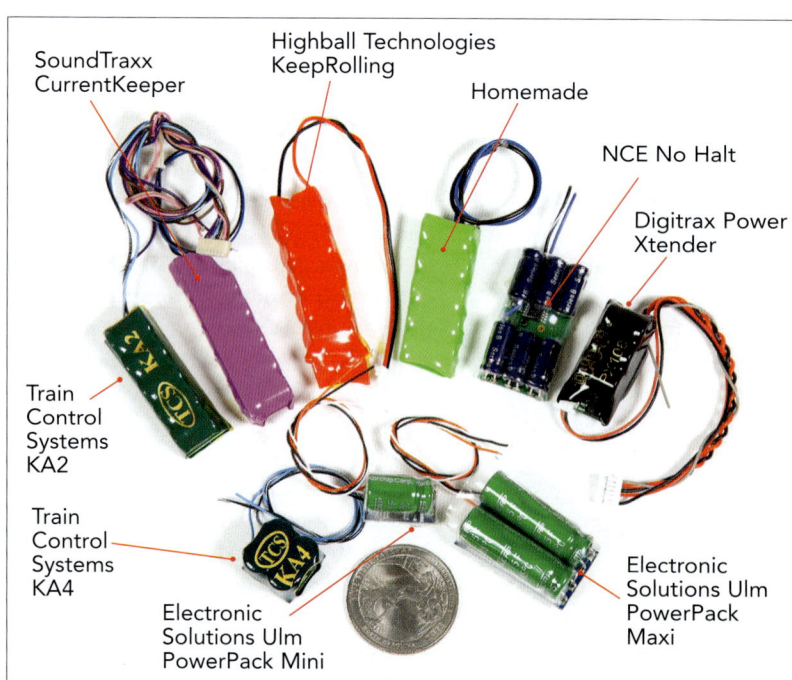

1. Stay-alive devices come in a variety of shapes and sizes, but most consist of four to five capacitors rated at 2.5V to 2.7V each, giving a potential total voltage of 10V to 13.5V.

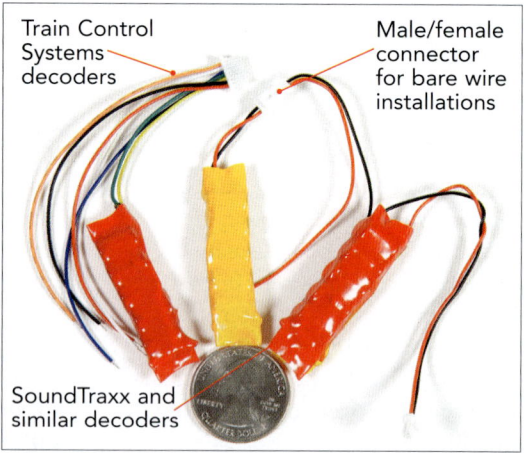

2. The new guy in town, Highball Technologies, offers several stay-alives, called KeepRolling, rated at 13.5V. They differ in the types of connectors. The bare wire device has a male and female connector in the middle of the wires to allow it to be easily removed for programming. These are available from Railmaster Hobbies (www.railmasterhobbies.com), as well as Bob the Train Guy (www.bobthetrainguy.com).

62

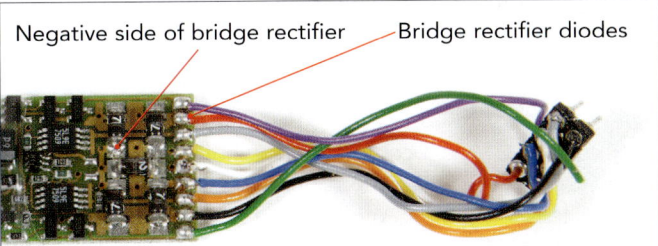
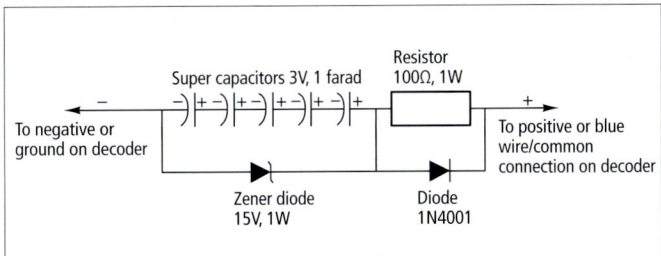

3. Four diodes on this Lenz LE080 mobile decoder form the bridge rectifier. The negative wire from a stay-alive should be connected to the negative output of the bridge rectifier and the positive wire to the blue common positive wire or connection point in the decoder.

4. This diagram shows the layout of components for a basic, stripped-down stay-alive circuit. Using third-party accessories may void the decoder warranty.

(www.members.optusnet.com.au/mainnorth/alive.htm). He has numerous photos of decoders showing the positive and negative contacts.

Now let's look under the wrapper at what's inside a stay-alive, **4**. First, there are anywhere from one to six capacitors —most now have four to five. Each of these super-capacitors is rated at 2.5V to 3V and 1F (Farad is a measure of how much current it can store). By connecting four to five of these in series, the device can store and release a total of 10V to 15V.

However, the voltage is controlled by the protection circuit in the stay-alive. If it's set at 13V, that's all you'll get out, even if you have a potential maximum of 15V. Also, if the decoder's bridge rectifier only puts out 12V, then your capacitors will only charge to that voltage, assuming the protection circuit allows it.

There are exceptions to these voltage limits. The Electronic Solutions Ulm (ESU) devices have one or two capacitors, yet still drive 12V motors with complex circuitry that boosts the voltage from the capacitor(s), but that comes at a price—shorter stay-alive run times.

Folks ask me, why can't you use just one or two capacitors? After all, you don't really need 10- to 20-second run times. Well, you need a minimum of 5V to 7V just to power the motor control, lighting, and sound circuits on the decoder. That means you need at least three of the 2.5V super capacitors if you don't have ESU's complex circuitry. Plus, you want a little extra to give you a little run time.

With the new 3V super capacitors (Mouser 581-SCCR12E105SRB), you could probably get by with three of them in the circuit, but that would require a more complex protection circuit. The downside is they're a little bigger than the 2.5V to 2.7V super capacitors.

So what else goes into a stay-alive? First, there's a resistor—about 100Ω, 1W—that prevents the inrush current from shutting down your booster when first powered up, **4**. There's a diode (1N4001) in parallel with the resistor that allows current to bypass the resistor when it's supplying power to the decoder. Most circuit diagrams also include a Zener diode (13.5V to 16V, 1W) that protects the capacitors if the input voltage exceeds a specific level.

Most commercial stay-alives include other components, but the one I described, and as shown in **4**, is a basic, stripped-down version. Space here precludes detailing the step-by-step process for making your own stay-alive, but I do have a video on my website (www.dccguy.com).

The circuit boards on the ESU PowerPacks, **5**, are clearly more complex than the circuit in **4**, which allows them to boost the voltage from only one or two 2.7V capacitors to the required level. These devices also have a circuit that bypasses the PowerPack during programming.

Another nice feature is the ability to determine the run time using CV113 in the LokSound decoder. A default value of 3 seconds gives more than enough power to get over most dirty spots and dead frogs. ESU

5. These ESU PowerPacks have regulated circuits that allow them to use only one to two capacitors. Note that they also have three wires instead of the usual two.

recommends limiting the run time to prevent locomotives from coasting through stopping blocks.

The one downside to stay-alives is they can interfere with programming, especially on a service-mode programming track. Most manufacturers recommend doing most programming before installing the stay-alive.

Another option is to use a quick connect plug so the stay-alive can be removed when programming. You can also try allowing the capacitors to charge on a powered track for a couple minutes before placing the locomotive on the programming track. I've never had problems when programming on the main.

Stay-alives provide a major improvement for DCC locomotive performance, but they're not a wholesale replacement for reliable trackwork, powered frogs, and clean track.

3 Installing DCC with sound in a brass diesel

1. Like all Alco PAs, Southern's PA-3s were a magnificent sight at the head of a passenger train, and the sound of the diesel engine was unmistakable. Larry Puckett used a SoundTraxx Econami sound decoder to bring the Alco 244 to life.

Brass locomotives can present different challenges compared to plastic models, but decoder installations usually aren't difficult. I'll go through the steps of installing a SoundTraxx Econami ECO-100 sound decoder in an HO scale Overland Models brass Alco PA-3 diesel, **1**. (The ECO-100 has since been discontinued—you can track them down online, or use an alternative such as the Tsunami2 TSU-1100 (wires) and the TSU-2200 with JST connector).

The ECO-100 is about the size of a standard N scale decoder and is rated at 1 amp, with four functions and a 1W, 8Ω amplifier.

The great thing about this decoder is the small size and 1 amp rating, which makes it compatible with a wide range of locomotive models. The sound files were chosen from among the most popular in the SoundTraxx Tsunami inventory. Some are new and others are remastered.

Engine sounds on the diesel version include the EMD 567 (non-turbo), 645 (turbo), and 710 (turbo); General Electric FDL-16 (modern); and Alco 244 (RS-3, FA, PA). A selection of 16 air horns and 6 bells are included, along with an array of other sounds, some user-selectable and others that play at random or defined intervals.

Brass locomotives are renowned for requiring tuneups before they'll run reliably, so before doing anything else, give the locomotive a test run. These models are often shipped unlubricated to make painting easier and to keep grease and oil from getting all over the locomotive during transport.

The journals on mine were never lubed at the factory, but a quick shot of Teflon grease on the gears and light oil on the axle and motor bearings took care of them. Although this model is about 20 years old, it has an efficient Mashima can motor and electrical pickup from the wheels on both sides of each truck.

Installation

The opening in the bottom of the body shell limited me to a component width of no more than ¾". I used a 1.1"-diameter Train Control Systems (TCS) speaker, which fit through the opening when turned sideways. The speaker is rated at 8Ω and 1W to match the decoder's amplifier.

To produce as much sound volume as possible I chose a SoundTraxx enclosure, **2**. These consist of a base, two rings, and a top, which when assembled in a stack with the speaker inserted came out to about .6" tall by about 1.2" diameter. Turned sideways, I could slip the whole package into the locomotive body.

Another neat feature of the Econami is the ability to install a SoundTraxx Current Keeper. These are wired in place of the small capacitor provided with the decoder. In this case, the blue wire on the Current Keeper goes to the blue wire on the decoder, and its black wire goes to the green-and-white-striped decoder wire. Make sure to install sections of heat-shrink tubing over the soldered connections.

For this locomotive I needed both a headlight and a Mars light. I was able to fit a 2.4mm, 14V Miniatronics bulb in the headlight casting, but had to use a smaller 1.7mm, 1.5V, 40 milliamp bulb in the Mars light. I secured both of these into the brass castings with cyanoacrylate adhesive (CA). The 1.5V bulb required a resistor to drop the lighting voltage down to 1.5V.

The Econami decoder's wired functions aren't regulated at 12V but instead are actually about 1.8V less than whatever your track voltage is (you can check your track voltage with a RRAmpmeter, available from www.dccspecialties.com).

To calculate the resistor size, all you need is Ohm's Law: $R=E/I$, where R is resistance in ohms, E is the voltage drop required, and I is operating current in amps. Since my track voltage is exactly 14.0V, $R=(14-1.8-1.5)/.04=267.5$, where 1.8V is the voltage drop from the decoder, and 1.5V is the drop from the bulb, and .04 is the amperage draw of the bulb.

When I tested the bulb with a 270Ω resistor, the light was too dim, so I went down to 220Ω and it was just right. Be sure to use heat-shrink tubing to protect the solder joints.

I wired the headlight to the white decoder wire and the Mars light to the FX3 brown decoder wire. The brown wire by default is assigned to function button F24, so I remapped it to button F3 and disabled the short horn. But that didn't work.

2. SoundTraxx enclosures consist of a base, two rings, and a top for securing the speaker. By adding or subtracting the stackable rings you can alter the finished height of the enclosure. For more bass and volume, make it as tall as possible.

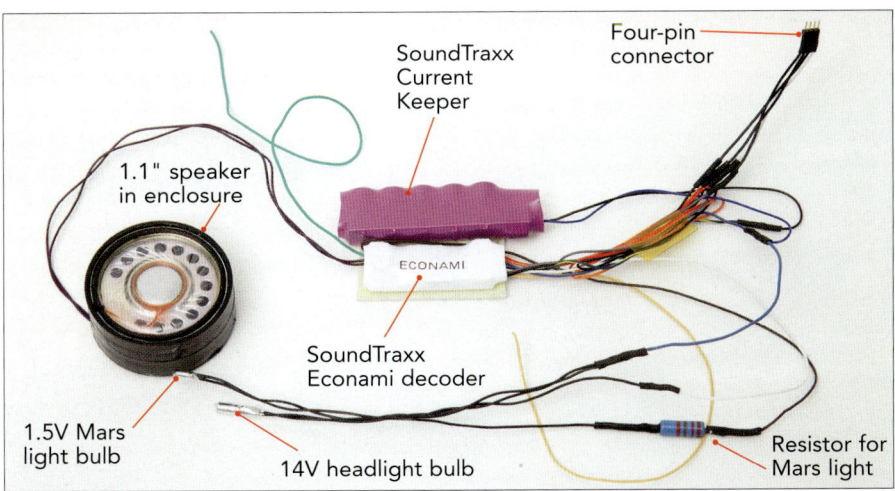

3. Larry assembled the speaker, decoder, Current Keeper, and lights as a unit, then slipped them into the opening in the body and secured the parts with double-sided foam tape.

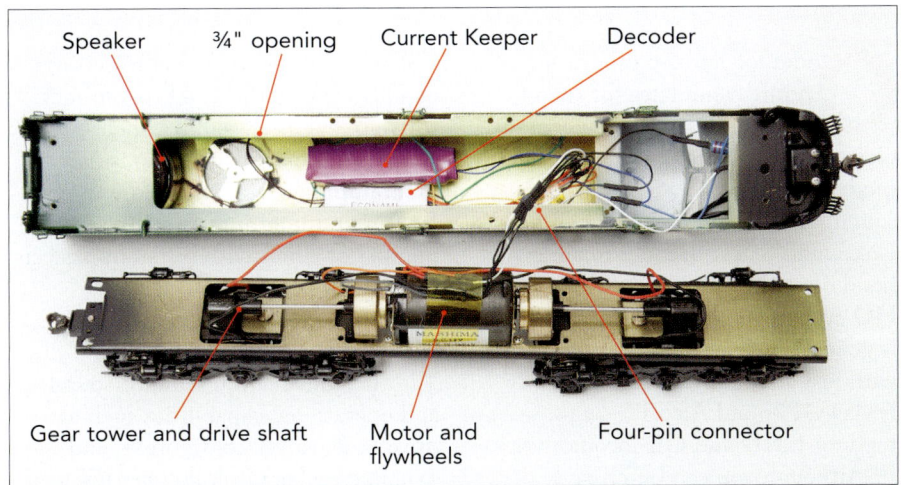

4. A Train Control Systems four-pin harness makes it easy to separate the body from the chassis for maintenance.

I subsequently learned a batch of ECO-100s got out of the factory with the green and brown wires reversed, so if you use FX3 and/or FX4 and they don't seem to work right, this may be why. Instead of rewiring the light bulb, I just remapped FX4 to F3 and programmed it accordingly.

All the above connections were hardwired to the decoder, **3**. However, the connections to the motor and track pickup wires required a different approach. I used a TCS four-pin harness for the connection between the motor and decoder. I wired the red, black, orange, and gray wires from the decoder to the connector plug, and repeated the same connections to the connector socket for the motor and truck wires.

This makes it easy to remove the locomotive body for maintenance. I installed the speaker, decoder, and Current Keeper in the locomotive using double-sided foam tape, **4**.

I made the final connection, reassembled the locomotive, and headed for the programming track. Why not give the locomotive a try on the main line with the default address 3? The programming track is limited to 250 milliamps, and is less likely to do any damage if you mis-wire something.

Also, many decoders today can detect a short and shut down before any real damage occurs. The Econami, for example, has a red error light-emitting diode (LED) that signals overloads and shuts down the decoder to prevent damage. (Note: the user manual says both the pilot and error lights are blue, but only the pilot light is blue).

Programming this decoder was easy, especially using DecoderPro. Most of the default values don't need changing, although I needed to configure the FX4 light as a Mars light and remap it to the F3 button (see page 86 for details on remapping decoder functions). Using the lighting panes and mapping function in DecoderPro, all it took was a few clicks and a write to the decoder for the change. Other simple changes included selecting the Alco 244 prime mover, Nathan M5 horn, and Alco bell.

So, how does it sound? Well, I'm an Alco fan, and I was blown away with that wonderful rattle of the 244 diesel engine. The speaker/enclosure combo delivers a nice level of bass frequencies, and the new electrical design of the Econami board, with its more powerful components, delivers a lot of volume. In addition, there's a 12 db boost on the horns, making them significantly louder as well.

For additional tips, and to hear the finished installation and see my Southern PA-3 in action, visit my website at www.dccguy.com.

3 Adding a LokSound decoder to an older diesel

1. Carolina & Northwestern Alco RS-11 no. 11 shoves a cut of cars onto the Chesapeake & Ohio interchange track in Charlottesville Yard on Larry Puckett's HO scale Piedmont Southern layout. The locomotive has an ESU LokSound Select decoder.

Older locomotive models are prime candidates for sound decoders, as they are generally nice, smooth-running, well-detailed models that simply lack DCC and sound. This is how I installed an ESU LokSound Select decoder in a high-hood diesel locomotive—an Atlas HO Alco RS-11, **1**. This installation is generic enough that it should work with just about any diesel from early EMD GP7s and Alco RS-3s to modern EMD and GE locomotives.

Although you can purchase a LokSound Select decoder with whatever available sound project you ask for, by using the LokProgrammer software and hardware you can download and install sound projects yourself. In addition, you can update them if new features are added.

There are two kinds of LokSound Select sound projects, called factory-equipped and retail. The factory-equipped projects are those ESU makes for model manufacturers for specific models. The retail projects are those in ESU's regular line. While either sound project version can be installed in a LokSound Select decoder, the factory-equipped versions usually have programming to meet a manufacturer's specifications.

Loading a sound project into a LokSound decoder is straightforward. You simply download the file from the LokSound website (www.projects.esu.eu), load it into the LokProgrammer software (Windows only), then send it to the decoder on an isolated programming track attached to the LokProgrammer interface. I tried it a number of times with different projects without an issue.

Many LokSound projects incorporate a new feature called Full Throttle. This feature set adds sound and operational capabilities such as drive hold, run 8, coast, brake, and dynamic brake. Look for the "FT" designation on a sound project with the Full Throttle features.

Installation

I used a LokSound Select Direct decoder for this project. It has a circuit board shape designed to be a direct replacement for the lighting circuit board similar to those made by Atlas and others. This direct replacement circuit board is the primary reason why I said this installation is generic enough to cover many locomotive models made during the last 30 or so years. Just to show you how easy this can be, I used one of the early RS-11 models Kato produced for Atlas in the mid-1980s.

To begin, I removed the body and plastic light board. I soldered wire leads to the motor brush contacts using the medium setting on my Weller WLC-100 soldering station. Next, I dropped the decoder in place on the mounting clips on top of the motor. There's small lettering on the board designating the front light (FL) and rear light (RL) to ensure correct orientation.

I soldered the motor leads to the contacts on the side of the board, and the left and right track pickup wires to their respective contacts on the board. At this point, I tested the installation on the programming track by changing the address. Afterward I gave it a test run.

Next I installed a speaker and lights. For this installation, I chose a Train Control Systems six-wire mini-connector harness, **2**. I like using these connectors, since they make it easy to remove the shell for maintenance and are color coded. I spliced yellow and blue wires in place of the red and orange wires. I soldered the ends of one set of wires to the proper solder pads on the LokSound board and attached the other half to the light-emitting diodes (LEDs) and speaker.

I matched the LokSound Select decoder with an ESU LokSound 1 watt, 4Ω speaker. This is a 16mm x 25mm speaker with a shallow plastic enclosure designed for use with LokSound decoders. The 16mm width provided a slide-in fit. I added a thin cardboard shim to wedge it in tight. I installed the speaker in the long hood end of the locomotive right behind the front LED, removing the front die-cast metal weight, **3**.

One nice feature of these boards is they have 2.2KΩ resistors installed on the function outputs. This allows direct installation of LEDs without the need for additional dropping resistors.

If you wish to use incandescent bulbs, there are pads for each function that can be jumped with a drop of solder to bypass the on-board resistors. However, if the bulbs are rated at less than 16V, you will need to add your own dropping resistors. This is all spelled out for Atlas, Athearn, and other locomotives on the instruction sheet that comes with the decoder.

In this case, I installed 2mm x 3mm surface mount device (SMD) N1022C super-incandescent white LEDs from Ngineering (www.ngineering.com). I explain how to install these in the project shown on page 74. The on-board resistor still allowed the LEDs to light brightly enough for my headlights. I shortened the factory-installed light tubes and attached the LEDs to them using cyanoacrylate adhesive (CA). To stabilize the wires, I attached them to the shell with Kapton tape, **3**.

Programming

After connecting the TCS harness, **4**, I slipped the shell on the chassis and moved to the LokProgrammer track. I downloaded the Atlas factory version of the Alco RS-11 sound project from the ESU website and loaded it in the decoder. If you're using a decoder that has a preloaded sound project, you'll want to change the address to match your locomotive, and you'll likely need to adjust the individual sound volumes.

You can also use Java Model Railroad Interface (JMRI) DecoderPro (www.jmri.sourceforge.net), or the standard programmer in your command station to change the address and other configuration variables (CVs).

However, the LokSound decoders do best with about 13V on the programming track, and some DCC system programming outputs are lower than that. If you have problems, try programming on the main, which uses full track voltage. Also, the indexed CVs used in LokSound decoders can create issues with some DCC system programmers, but work-arounds are explained in the manual.

The master sound volume is controlled with CV63. However, because these decoders use indexed CVs for individual sound volumes, you'll need some extra steps for them. In each case below, program CV31 to a value of 16 and CV32 to a value of 1 before entering the desired CV and its value. The prime mover volume would require changing CV259, the horn is CV275, and the bell is CV283. The range is between 0 and 128.

Other programming changes can get a bit more complicated, so you'll need to refer to LokSound and other documentation. To make it easier to find what you need, I've placed several documents on my website (www.dccguy.com) for easy downloading. You can also find additional information on the LokSound website (www.esu.eu/en/start/) and its Yahoo group. Documentation specific to a particular prototype sound project can also be found in the project description on the LokSound download page.

The sounds from my RS-11 replicating an Alco 251B diesel engine are exactly what modelers have come to expect from a LokSound decoder—just like being trackside. Motor control is equally impressive. The fact that you can install the same sound projects in your existing locomotives gives you the ability to match both sounds and functions with factory-installed decoders in newer models, an important consideration if you want to use the new Full Throttle features.

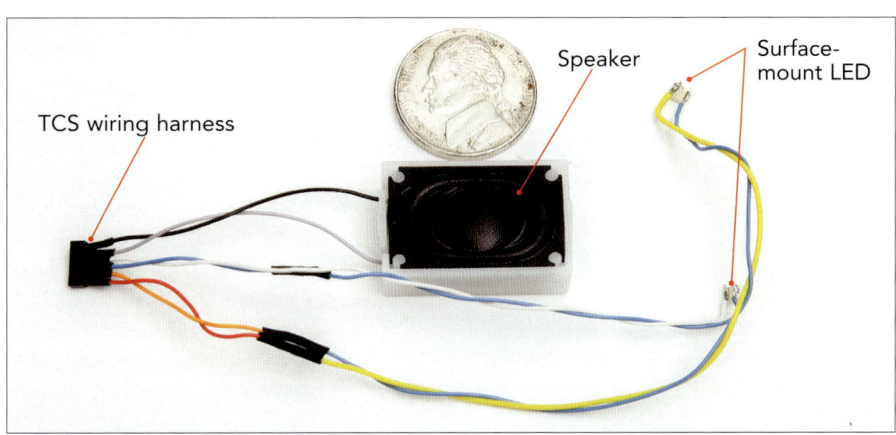

2. Larry used this ESU Loksound 1 watt, 4Ω speaker and a Train Control Systems six-wire harness for this project.

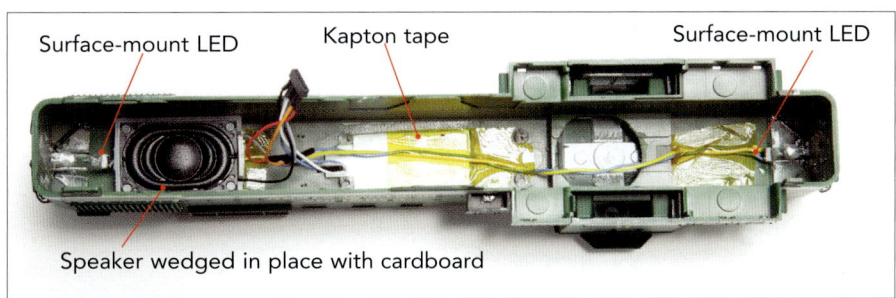

3. After tucking everything into the locomotive shell and securing the wires with Kapton tape, Larry slid the speaker into place.

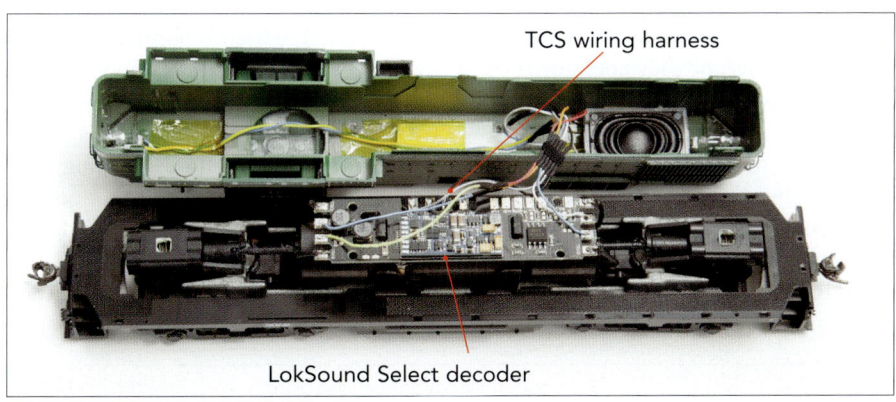

4. With everything installed it was a simple matter to connect the TCS wiring harness and slip the shell on the chassis.

3 Dual engines with a WOWSound decoder

1. Southern E6A no. 2800 heads south with *The Southerner* on Larry Puckett's Piedmont Southern railroad. Larry shows how to get the sound of an E unit's dual diesel engines running in and out of sync with one another using a TCS diesel WOWSound decoder.

An interesting feature of Electro-Motive E-units was their use of dual diesel engines to get the horsepower needed for fast acceleration of long passenger trains. Their two 12-cylinder 567 diesel engines were sometimes synced and sound like one loud locomotive, or they could drift out of sync a touch and sound like a pair of locomotives running together.

Short of installing two sound decoders, how do you duplicate this with your models? Train Control Systems (TCS) has provided a way to make it sound as if two engines are growling away in your favorite E unit, and you can control how far out of sync they are.

Installation

I installed a WOWSound version 4 decoder in an older Life-Like Proto 2000 E6A, **1**, which is similar to the recent WalthersProto DC models. The Walthers model is easier to work with since it has a 9-pin socket, as well as holes in the standard 8-pin plug arrangement for an 8-pin socket or for soldering the wires from the decoder.

With the older Life-Like model there are two approaches, the easiest being to add an 8-pin plug to the decoder and plug it into the socket in the model. The downside to this approach is finding a place to put both the decoder and the speaker. Although there's a slot in the weight for a decoder, the WOWSound decoders are a bit too long to fit.

I chose the second approach, removing the locomotive's circuit board and hard-wiring the decoder in its place. After I removed the screws that hold the circuit board in place, I marked all the wires with masking tape, then disconnected them.

Life-Like wasn't all that particular about the color of the wires used, and the colors generally don't match the National Model Railroad Association (NMRA) standard (see the chart on page 69). Fortunately Life-Like did mark the "+" and "-" motor wires on the circuit board.

I cut a rectangle of black styrene sheet sized to fit in place of the circuit board and attached it to the chassis with double-sided foam tape. I used the same tape to attach the decoder and capacitor to the styrene sheet, **2**. Because of the model's heavy weight, 12-wheel electrical pickup, and long wheelbase, a TCS Keep-Alive capacitor circuit isn't necessary.

Since I was mimicking the sound of dual diesel engines, I installed dual speakers (which were probably a bit of overkill, but read on). I chose a pair of TCS WOWSound sugar cube speakers (no. UNIV-SH5-C). The 9mm x 16mm, 8Ω, 0.7 watt speakers put out 97 dB each, **3**. I wired these in series giving 16Ω impedance, the preferred WOWSound configuration. (For more on speaker choices and installation see page 60.) The speakers are available with a 3-D-printed enclosure that provides an excellent fit. The small spring clips on the speaker serve as solder points.

I marked the holes and also the wire contacts on the speaker as "+" and "-" to prevent any wiring mixups. Their actual polarity isn't marked and it really doesn't matter as long as you're consistent.

For series wiring, one purple wire from the decoder is attached to the "+"

2. Larry installed a piece of black styrene in place of the original circuit board, then attached the WOWSound decoder, capacitor, and speaker to it using double-sided foam tape. It's a good idea to paint the side of the foam tape black so it won't be visible through the side window of the locomotive.

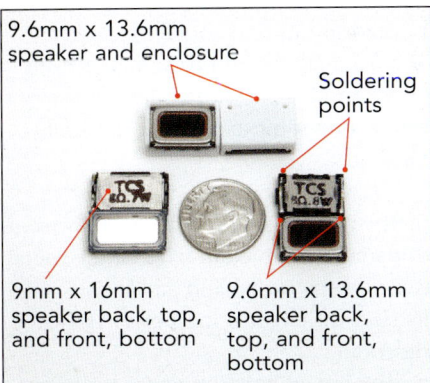

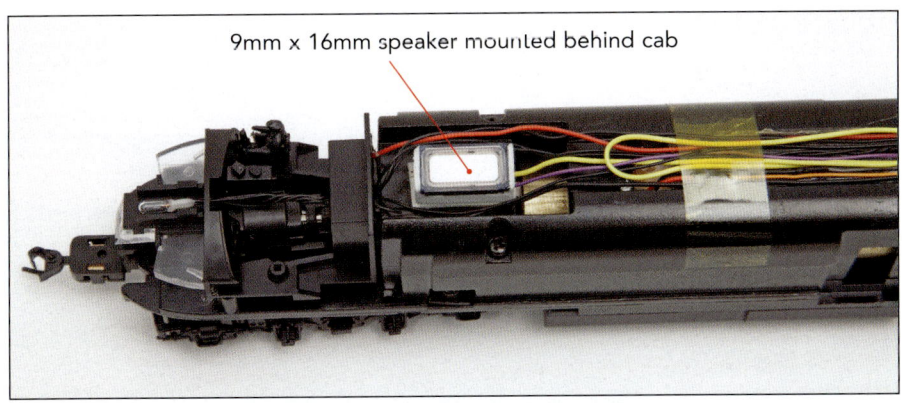

3. TCS offers two sugar cube speakers, shown here. The 9mm x 16mm version has a much lower 300 Hz frequency than the 9.6mm x 13.6mm one at 700 Hz. The plastic enclosures are necessary for good sound output.

4. Larry placed a second speaker in the slot behind the locomotive cab. In hindsight, one speaker probably would have been adequate.

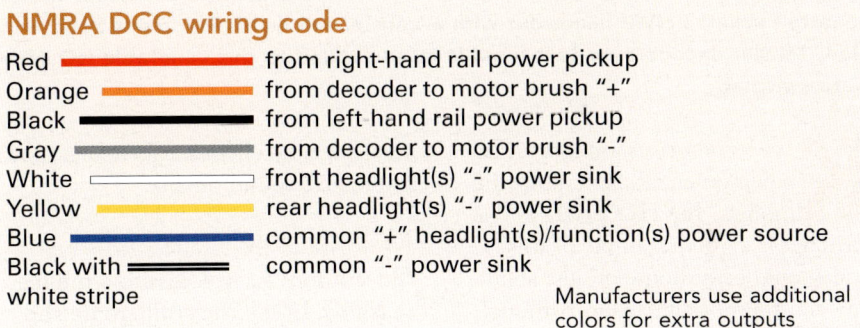

contact on one speaker, and the other purple wire goes to the "-" contact on the other speaker. I then ran a yellow wire between the other two contacts. In hindsight, one speaker would've been enough.

Although the enclosures provide a good fit, they still need to be sealed for the best sound. I used gap-filling cyanoacrylate adhesive (CA) applied to the gap between the speaker case and the enclosure. Don't get CA on the speaker diaphragm (the flexible joint between the speaker cone and frame). I installed the speakers using double-sided foam tape, with one on the styrene sheet behind the decoder, **2**, and the other in the slot behind the cab, **4**.

Decoder settings

There are two settings for sound synchronization provided in WOWSound diesel decoders. First there's a configuration variable (CV) setting to turn on the dual prime mover option, and a second which sets the notch-up time delay for the two diesel engines. Both are controlled using indexed CVs 201, 202, 203, and 204.

Calculating the values for these settings requires choosing from among several options and adding up their individual digital values. These are in a table in the Diesel Programming Guide available as a download on the TCS website. Look for the table under User Options. The time delay programming settings are in a second table under Dual Prime Mover Delay toward the end of the manual.

If calculating digital CV values isn't for you, go to the TCS website and use the online WOWSound Guided Programmer (www.tcsdcc.com/Customer_Content/Technical_Info/WOWSound/Diesel/guided_programming_diesel_vers4.php) to get the values. Make sure to use the programmer for the correct decoder version. If you use Java Model Railroad Interface's DecoderPro, it will calculate and enter the values for you.

Activating the dual diesel engine feature is a simple on or off selection, but the delay factor requires a little more thought. The time delay is calculated as CV value/2. A value of 1 will result in a notch up delay of ½ second, and a value of 4 will result in a 2 second delay. The maximum value is 20 and the default is 5. The final result will be the sound of two engines notching up as you increase the throttle setting, with the chosen time delay separating them. For models, a 1 to 2 second delay seems realistic.

I replaced the factory-installed bulbs with Miniatronics 14V 30mA bulbs (no. 18-014-10). With 14V on the DCC tracks, the bulbs only see about 12V. I wired the headlight to the blue and white decoder wires to operate on function 0. For the Mars light I wired the bulb to the pink and blue wires for F4 activation. I used DecoderPro to program the lights.

With all the wiring and programming done, I popped on the shell and gave the E6 a test run. The dual speakers are a bit of overkill. But they do create a stereo effect that seems to enhance the volume—a lot. Hopefully my description of the programming options will be enough to get you started. But for more details, visit the TCS website (www.tcsdcc.com) and my website (www.dccguy.com).

3 Packing sound into a vintage HO Kato switcher

1. Larry Puckett's NW2 heads out with a transfer run on his Piedmont Southern. New, smaller decoders made it feasible to add sound to this venerable HO scale Kato switcher.

Kato locomotives are known for being quiet, efficient, powerful models. The HO EMD NW2 switcher is a good example of this, **1**, and should be at home on just about any era-appropriate layout. Although they've been out of production for a few years, there are thousands of them floating around at train shows, swap meets, and internet auctions.

My NW2 is a smooth and reliable runner that can out-pull all of my other switchers. Part of the reason for all the power is its weight, which checks in at a hefty 12.5 ounces. However, all that power and weight come at a cost—there isn't room under the hood for even a stray hair!

Kato used a split die-cast metal frame that completely fills the shell. The cab's detailed plastic interior further complicates a sound decoder installation. However, recent developments in miniaturization convinced me it was time to give sound a try. You can use similar techniques on other older switcher models as well.

Initially, I was challenged to find a place for the decoder. I didn't want to sacrifice the cab interior, so I looked for a way to add a decoder under the hood. Because of the tight fit, my only option appeared to be cutting out a space in the chassis. To keep the cutting to a minimum, I selected an ESU LokSound Select Micro decoder for its small size, excellent performance, and impressive sound quality.

Making it fit

Since I don't have access to a milling machine, I took the crude route and hacked out a rectangle from the front of one half of the die-cast chassis, **2**. This was actually easier than it sounds, using a Dremel EZ Lock metal cutting wheel.

One cut across the chassis and another lengthwise created a space large enough for the decoder and wires. I made my cuts at the front of the chassis to avoid the motor and flywheel farther back. To make room for the wires going to the cab, I also cut a V-shaped notch in the top of the chassis, **2**.

With the cutting out of the way, I moved on to the motor. Power is picked up from the frame halves using metal wipers attached to the motor brushes. I clipped these wipers short, leaving enough for solder pads, and attached orange and gray wires to the bottom and top contacts, respectively. I fed the orange wire up through the plastic frame that insulates the motor from the chassis and fed both wires out through one of the small cast-in holes, seen in **3**.

Next was providing power pickup. Since each half of the frame was designed to conduct track current, I opted to tap that source of power. The original circuit board at the front of the model slides into slots in the frame, providing power for the LED headlight.

I removed the LED and resistor, then cut the board in half lengthwise. I soldered a red pickup wire to the metal trace on the board and slid it back into its slot, which gave me power from the right rail.

The left rail pickup was a bit more difficult. Since I had cut out the front section of the frame on the left side, I couldn't use the circuit board trick there. I drilled a .020" diameter hole in the frame next to the decoder and inserted a piece of .020" brass wire. I was then able to solder the black wire from the decoder directly to this wire.

I didn't want to reuse the old LED headlight, as it was large and had an orange tint. I ordered some small surface-mount device (SMD) LEDs on eBay from a supplier in China—many sources offer them. I chose versions with the wires already attached. I ordered both warm white and cool white versions.

The warm white was too orange and the cool white had a blue tint. Using an old trick, I coated the cool white LEDs with Tamiya X-26 clear orange paint. A thin coat corrects the color and gives a pleasant golden white light.

Using cyanoacrylate adhesive (CA), I attached a small surface mount resistor (1,000Ω, ½ watt) to the underside of the old circuit board I used for power pickup. I then soldered the blue wire from the decoder to one

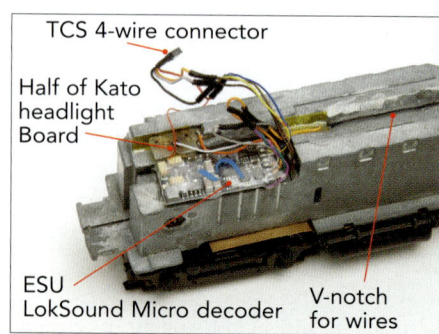

2. Larry cut a rectangle from the front half of the die-cast metal chassis for the decoder and added a V-notch on the top of the chassis for the wires.

70

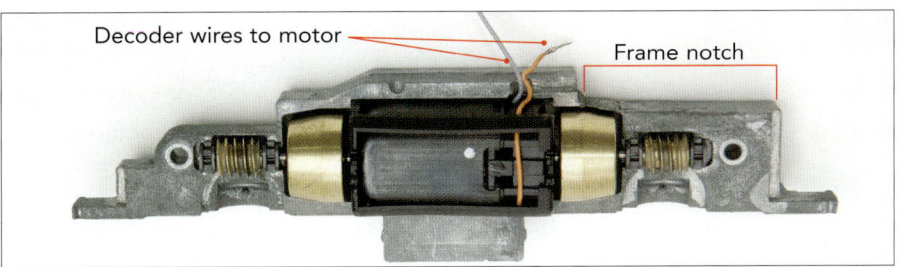
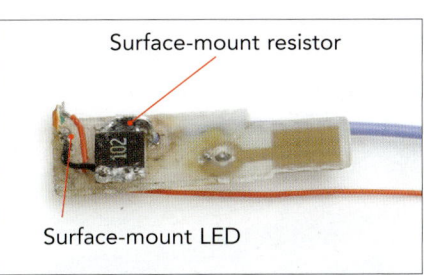

3. Larry fed the orange and gray wires up through the plastic frame that insulates the motor from the chassis and out through one of the small cast-in holes. He cut the notch in the frame for the ESU LokSound Select Micro decoder with a metal-cutting wheel in a motor tool.

4. Larry glued a 1KΩ surface-mount resistor to the underside of the old light board and attached the wires from the decoder.

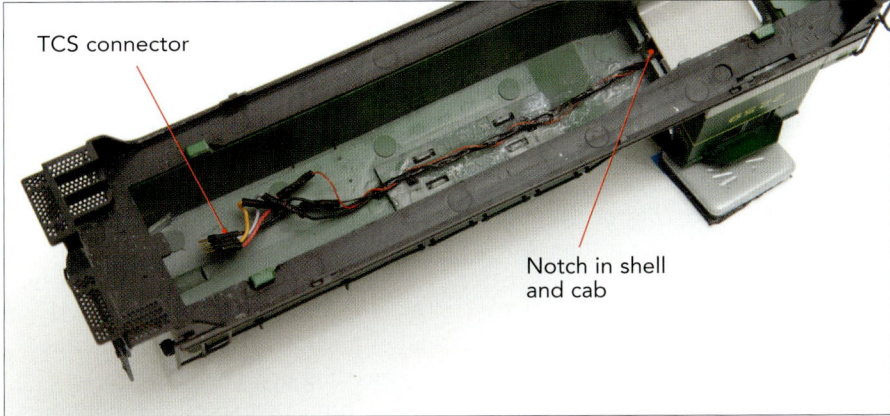

5. In a similar fashion, Larry glued a 1KΩ SMD resistor to the side of the sugar cube speaker enclosure and attached the LED to it.

6. After installing the speaker and light into the roof of the cab, Larry then glued the wires to the cab inside the shell, making sure they were perfectly centered to match the V-notch he had cut into the top of the frame. The TCS connector simplifies locomotive service by making the shell removable.

side of the resistor and the red (+) wire from the LED to the other side.

Finally, I connected the black (-) wire from the LED to the white wire from the decoder, **3**. This placed the resistor and LED under the circuit board when installed and lined up with the back side of the headlight insert.

Sound

I removed the old headlight enclosure in the cab to make room for a sugar cube speaker. An 11 x 15 mm speaker from Streamlined Backshop appeared to be a good fit, and being all black, it wasn't too visible through the cab windows. For the headlight, I used another small LED with an SMD resistor cemented to the back of the speaker enclosure, **5**.

A small piece of double-sided foam tape secured the speaker to the roof of the cab with the LED sitting conveniently inside the recess in the headlight. Painting the flanges of the headlight casting and plugging the opening with some putty will prevent the cab being lit up whenever the rear headlight is on.

After installing the windows in the cab, I cut a notch in the front of the cab insert and a matching one in the rear of the shell for the wires to pass through. While centering the wires in the front of the cab, I installed the cab detail insert, then mated it to the shell. I glued the wires inside the shell, making sure they were centered, **6**, to match the V-notch I'd cut into the top of the frame.

I used a TCS 4-wire connector for the shell wiring, **2**. Since most of the connector wire colors didn't match those from the speaker and headlight wires, I just kept track as I soldered them. With the wires all connected, I mated the male and female ends and slid the shell onto the frame, then moved to the programming track.

I downloaded and installed the correct EMD 567 prime mover sound file from the LokSound website. The LokProgrammer interface and program makes this a quick and easy task. Here's a hint: If you buy a LokSound universal decoder, make sure you ask the dealer to install the correct prime mover sound file for you. A friend installed a LokSound decoder in his steam locomotive and when he turned it on, only diesel sounds came out.

I prefer to do the bulk of my programming using DecoderPro in the Java Model Railroad Interface (JMRI, www.jmri.org), so I added the locomotive to the roster and programmed the address. For a complete introduction to using DecoderPro, see the video tutorial series on my website (www.dccguy.com).

The LokSound default settings typically require few changes other than the address, so I was running the switcher in just a few minutes.

3. Back-EMF and a quick decoder installation

1. Southern F7A no. 6718 leads a consist on Larry Puckett's Piedmont Southern layout. Larry describes back-electromotive-force control and installs a decoder in the HO scale Walthers locomotive.

Back-EMF, or back-electromotive-force control, is now a common decoder feature. But just what is back-EMF, how does it work, and what are its advantages? First, let's look at the basics behind back-EMF.

DCC decoders power motors, **1**, using a method called pulse-width modulation (PWM). With PWM, a series of full voltage pulses are sent to the motor, making it turn a little with each pulse. The longer the pulses are on (the width), the faster the motor will spin. Send out a series of short (narrow) pulses and the locomotive will move slowly. Increase the duration (width) of the pulses and the locomotive will speed up proportionally.

What does PWM have to do with back-EMF? Direct-current motors turn when an electrical current is applied to their windings. If you spin a DC motor by hand it will generate a small current.

With PWM, the motor is constantly spinning, but there are periods between the pulses where no current is being applied, **2**. During each of those gaps between the pulses, the motor generates a brief pulse of current that travels back to the decoder. That's the back-EMF.

How it's used

So how does a decoder use that back-EMF? The current generated will be proportional to the speed at which the motor is turning, **2**. Using that relationship, the decoder senses not only how fast the motor is turning, but by comparing a series of back-EMF pulses it can also determine whether it's speeding up, slowing down, or maintaining a set speed. By monitoring back-EMF, the decoder can adjust the pulse widths it's sending to the motor and control the motor's speed. One nickname for back-EMF is cruise control for model locomotives. With back-EMF working, you can throttle up to a desired speed and the decoder will use back-EMF to maintain that same speed even going up or down grades.

To achieve this level of control requires some complex mathematics. One commonly used approach is called a PID controller. This approach uses three factors: Proportional, Integral, and Derivative, to make the constant speed adjustments, **3**. Each factor monitors a different aspect of motor response, and their combined adjustments keep the motor turning to provide a constant speed even under varying loads.

Not all decoders use the PID approach. Some may use fewer factors, while others may use a completely different method. For example, SoundTraxx only uses the P and I factors. However, if there are three adjustment variables described in your decoder manual, then it's likely they're using the full PID method.

Adjusting these variables can get quite complicated. Both Train Control Systems and Electronic Solutions Ulm (ESU) LokSound decoders have automatic calibration procedures to help simplify this process. However, I've never felt the need to make adjustments to the factory programmed values in the currently available decoders I've used.

Because the decoder is constantly monitoring the back-EMF and adjusting the pulses to maintain a constant speed, it can provide very smooth operation, especially at slow speeds. Once the locomotive starts to pick up speed, the usefulness of back-EMF decreases. Therefore, some decoders cut out back-EMF above a certain speed step.

Consisting and sound

Back-EMF control works great as long as you're using a locomotive by itself, but complications may arise with consist operation. If the locomotives in the consist aren't speed matched (see page 90), then they'll fight one another.

So the first step with consist building is to speed match all the locomotives involved. Also, do the automatic calibration procedure if you're using the TCS and LokSound decoders. If things still aren't smooth, then do an internet search—there are numerous websites that offer help with adjusting the control variables in different decoders. In the end you may actually find it's easier to consist locomotives of the same type and to use the same decoder type in all locomotives in a consist.

Back-EMF is also used in sound decoders. By monitoring the back-EMF pulses, the decoder can sense how hard the locomotive is working and adjust related sounds in response. For example, most sound decoders now use back-EMF to increase the volume and cadence of the chuff of a steam locomotive when the decoder senses the locomotive is working harder.

Diesel decoders respond similarly, increasing the volume of the prime mover and changing the rpm level.

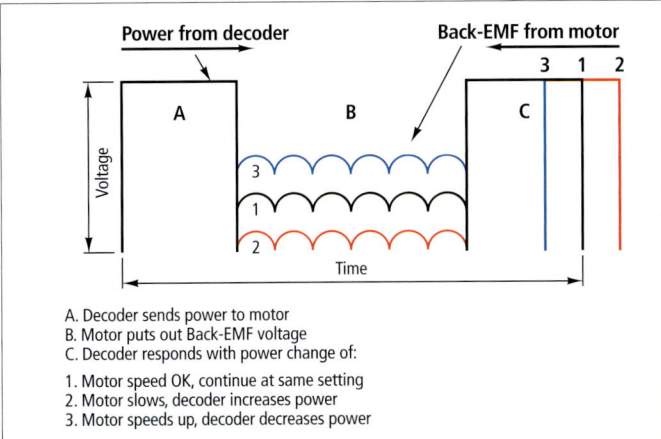
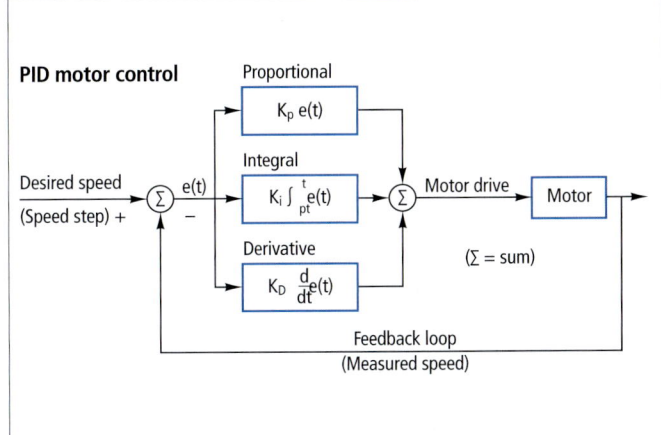

2. Back-EMF control uses the current generated by the motor between the DCC power pulses to monitor how fast the motor is turning. Depending on how much current is generated, the decoder can detect the speed of the locomotive and make speed adjustments accordingly.
Diagram courtesy of American Hobby Distributors

3. Many decoders use the Proportional, Integral, and Derivative (PID) control method to maintain a constant speed. By monitoring the back-EMF current generated by the DC motor between DCC pulses, the decoder uses advanced algorithms to maintain a constant speed.
Diagram used with permission of Mark Gurries

Back-EMF control is so integrated into TCS WOWSound decoders that the company doesn't offer the option of turning it off.

Walthers Mainline F7A

I recently acquired a trio of Walthers Mainline HO scale F7A locomotives, including one without a DCC decoder, so let's take a quick look at adding DCC. The DCC version of this model comes with a SoundTraxx Tsunami-based decoder, and the DCC-ready version comes with a printed-circuit (PC) board designed by SoundTraxx.

This PC board offers two approaches for adding a decoder. At the rear of the model there's a small circuit board attached with a 9-pin JST connector to wires running to the main PC board. It's a simple procedure to disconnect this board from the wires and attach a decoder in its place.

Since I wanted to use a SoundTraxx sound decoder to see how well it would operate with the factory-installed SoundTraxx decoders, I used an Econami ECO-100 I had on hand. Because these decoders don't have a 9-pin JST connector, this gave me the opportunity to test the other conversion method—soldering the wires into the eight holes on the PC board. (Alternatives to the ECO-100 include the Tsunami2 TSU-1100 with wires or TSU-2200 with a JST connector).

The PC board is relatively uncluttered, so I soldered the wires in the holes without removing the PC board. Although the Walthers instructions didn't say anything about removing the small board with its JST connection in the rear, I pulled it just to be safe, in case it acts as a shorting plug for DC operation. I then added a sugar cube speaker in the space over the rear truck where the JST connector board had been, **4**.

The reassembled model ran smoothly with the new sound decoder. And because I used a SoundTraxx decoder, it was perfectly matched with the factory-installed SoundTraxx Tsunami decoders.

When consisted, there was no lurching or push-me pull-you fighting due to the back-EMF-based Hyperdrive2 control algorithm in the decoders. Even though the Econami and older Tsunami decoders are a generation apart, the control approach and variable settings are so well matched I didn't need to make any adjustments to the back-EMF settings or speed curve.

Back-EMF control is a useful component of DCC decoder design and conflicts can be avoided, including by installing matching decoders in consisted locomotives.

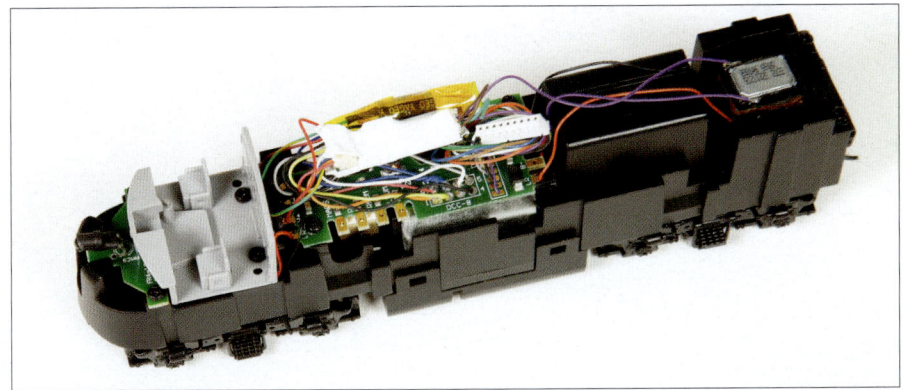

4. The circuit board in the Walthers DCC-ready Mainline F7A units, designed by SoundTraxx, makes converting these models to DCC an easy process. You have the option of either installing a decoder with a 9-pin JST connector or soldering the wires into eight holes on the main board, as shown here.

3 Squeezing sound into an HO SW1 switcher

1. A pair of Walthers SW1 switchers picks its way into Charlottesville Yard with a string of cars on Larry Puckett's HO scale Piedmont Southern layout. Larry Puckett added DCC and and sound to these compact models.

Back in 1993 when Walthers first released its model of an Electro-Motive SW1 switcher, DCC with sound wasn't an option. Then for many years squeezing a large sound decoder and speakers into a small locomotive was difficult or impractical. However, in the past few years, advances in speaker and decoder miniaturization, along with several reader requests, convinced me it was time to give it a try.

What made it possible? First came sugar cube speakers. These little devices range in size from about 8mm x 12mm to 13mm x 18mm and larger, and produce excellent sound for their size. Then SoundTraxx released the Econami ECO-100 micro sound decoder, which is about the same size as the N scale decoder I'd previously used. (Although the ECO-100 is no longer made, alternatives include the Tsunami2 TSU-1100 and TSU-2200.)

I initially planned to install an ECO-100 decoder in my vintage 1993 SW1. However, after a few quick measurements, I realized my only option would be to install the decoder in the cab. The Mashima flat can motor Walthers used in the model was just too tall for much of anything else to fit under the hood. And although I might've been able to fit a sugar cube speaker in the cab as well,

2. Larry's friend Doug Miller milled a space in the fuel tank of the early SW1 big enough for a small speaker.

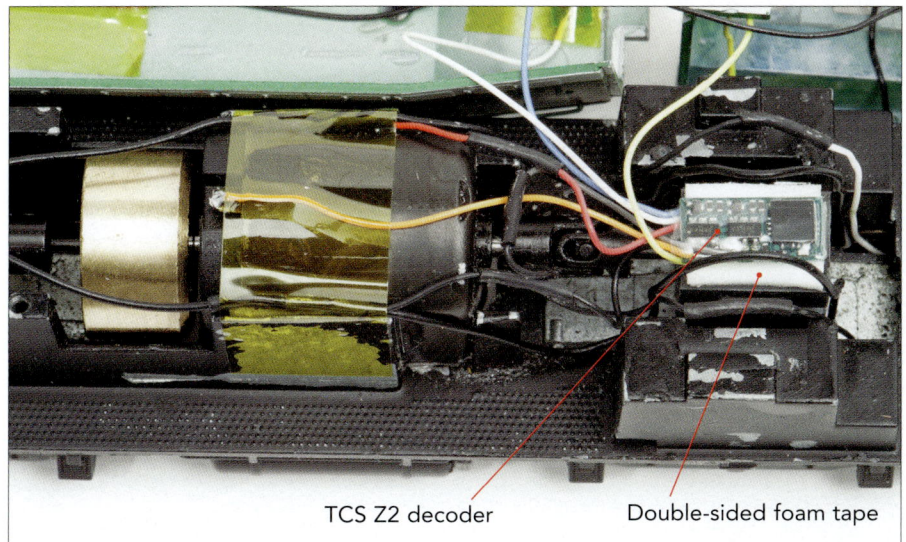
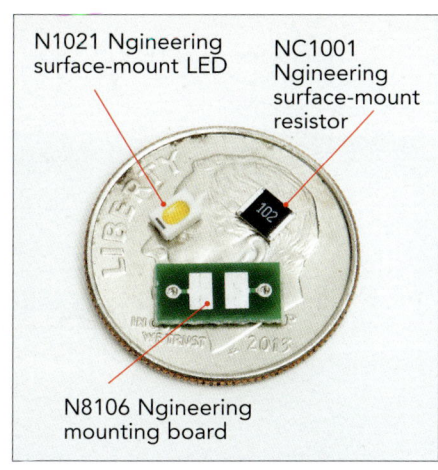

3. A tiny TCS Z2 decoder powers the Mashima can motor and lights in the early Walthers SW1 switcher.

4. These surface-mount components make installing the lights much easier and produce less heat than incandescent bulbs and full-sized resistors.

it likely would've been a troublesome installation, so I started to look for options.

I discovered that in its 2014 release of the SW1, Walthers used a much smaller motor, which also allowed much shorter truck gear towers. I quickly ordered a newer model, and after it arrived I found that both an ECO-100 and sugar cube speaker would easily fit under the hood.

At that point I had both the old and the new models, which left me with the conundrum of what to do with the old SW1. I finally hit upon a solution. The Southern Ry. had several SW1 locomotives with multiple-unit capability, so I decided to operate these two models that way, **1**. Let's go through the steps of installing decoders and speakers in each locomotive.

Installation

On the old model, my first move was to cut a chunk out of the fuel tank big enough to fit the speaker facing down toward the track. I used a 13mm x 18mm TDS SuperSonic speaker from Tony's Train Exchange, securing it in the opening with double-sided foam tape, **2**.

Because the TDS speaker is recessed into its enclosure, it has a slightly shorter profile than some of the other 13mm x 18mm sugar cube speakers I've used. I soldered two wires to the contacts on the speaker and ran them up the end of the motor, securing them with cyanoacrylate adhesive (CA).

For the decoder, I selected a TCS Z2, a Z scale decoder about the size of my little fingernail, yet rated at 1 amp. No, it doesn't have sound, but we'll get to that in a bit.

I soldered the orange and gray wires to the motor contacts and the red and black wires to the right and left pickup wires, protecting the joints with heat-shrink tubing. A small square of double-sided foam tape holds the little decoder firmly in place on top of the rear truck as shown in **3**.

Since I was rewiring the model, I replaced the old lightbulbs with more-efficient and longer-lasting LEDs. However, instead of my usual 3 mm golden-white LEDs, I tried a new product, surface-mount LEDs from Ngineering (www.Ngineering.com).

These neat little devices come in bright white (N1021) and golden incandescent (N1022C) versions, plus there are even smaller micro and nano versions for tighter applications. For my HO scale installation, I ordered the company's standard size golden incandescent version, **4**.

The trick to soldering wires to miniature components like these is to find a way to firmly hold them during the process. My solution was to attach a piece of double-sided tape to a flat surface, then place the LED on it bottom-side-up to hold it.

The application note on the Ngineering website suggests laying the wire across the solder pad to the LED, soldering it in place, then bending it up. After attaching the wires, I glued the LEDs to the rear of the lenses using CA and held the wires in place with Kapton tape, **5**.

To drop the current down to an acceptable level for the LEDs, I needed to add a resistor on the negative leads. Using the calculators on the Ngineering website, it said I only needed a 440Ω, ¼ watt resistor. However, I found this far too bright, and doubled the resistance using a 1KΩ, ½ watt surface mount resistor (NC1001), **4**.

This combination not only gives me more than enough brightness, it means the LEDs will last the rest of my lifetime and the resistor will not even feel warm when in use.

Since I was experimenting with surface-mount devices, I decided to use another Ngineering product. Along with a handful of 1KΩ, ½ watt surface mount resistors, I ordered mounting boards (N8106). The latter are small circuit boards for mounting the resistors and attaching wires, **4**.

I stabilized the board using the double-sided tape trick and attached the surface-mount resistors with a tiny drop of CA, then soldered the resistors

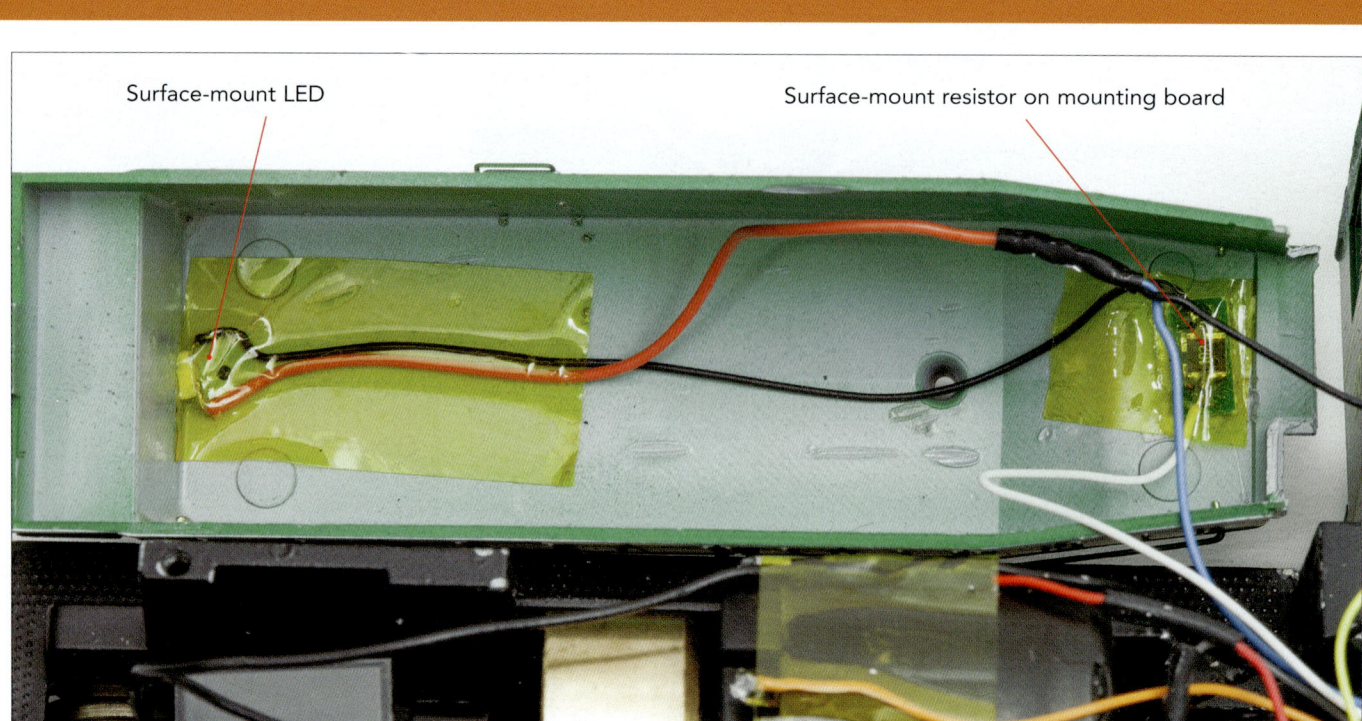

5. The surface-mount resistor circuit boards are attached to the inside of the shell with Kapton tape. Larry cemented the LED to the back of the headlight lens, then used Kapton tape to stabilize the wires.

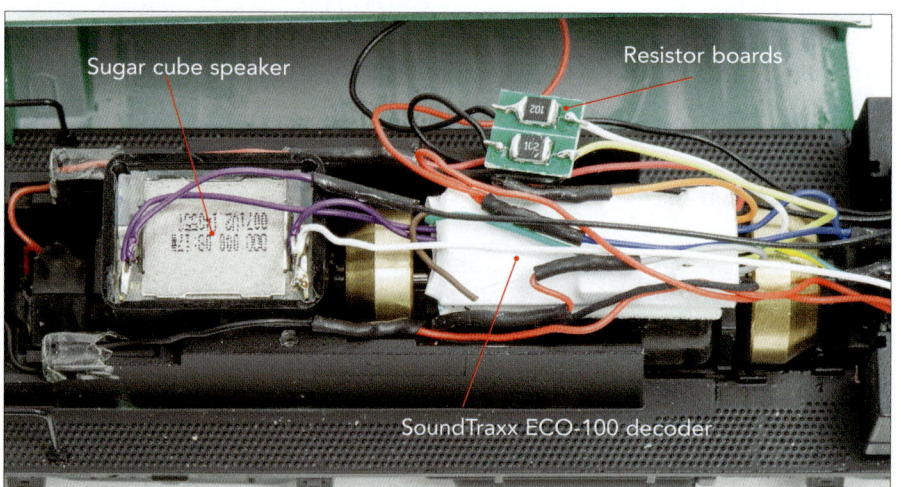

6. Larry installed an ECO-100 decoder in the new SW1 switcher to power it and to provide sound in both locomotives. In this case he left the surface mount resistors sitting on top of the decoder.

to the mounting pads. Finally, I attached the white and yellow decoder wires to the forward and rear light negative wires by connecting them to the resistor mounting boards and attached these to the inside of the shell, **5**. I then soldered the blue wire to the two positive LED wires.

Newer model

Next, I moved to the 2014 model. To prevent mixups, I marked the various wires before removing the existing circuit board. The wires are all either red or black, and some on the trucks are reversed, so you can't go by color alone. Pay special attention to the polarity of the wires that go to the LED headlights.

After removing the circuit board, I installed the ECO-100 decoder on top of the motor using double-sided foam tape, **6**. I then connected the orange and gray motor leads and the red and black track pickup wires. As with the old SW1, I made up the resistor boards and soldered the white and yellow wires to them, then connected the negative

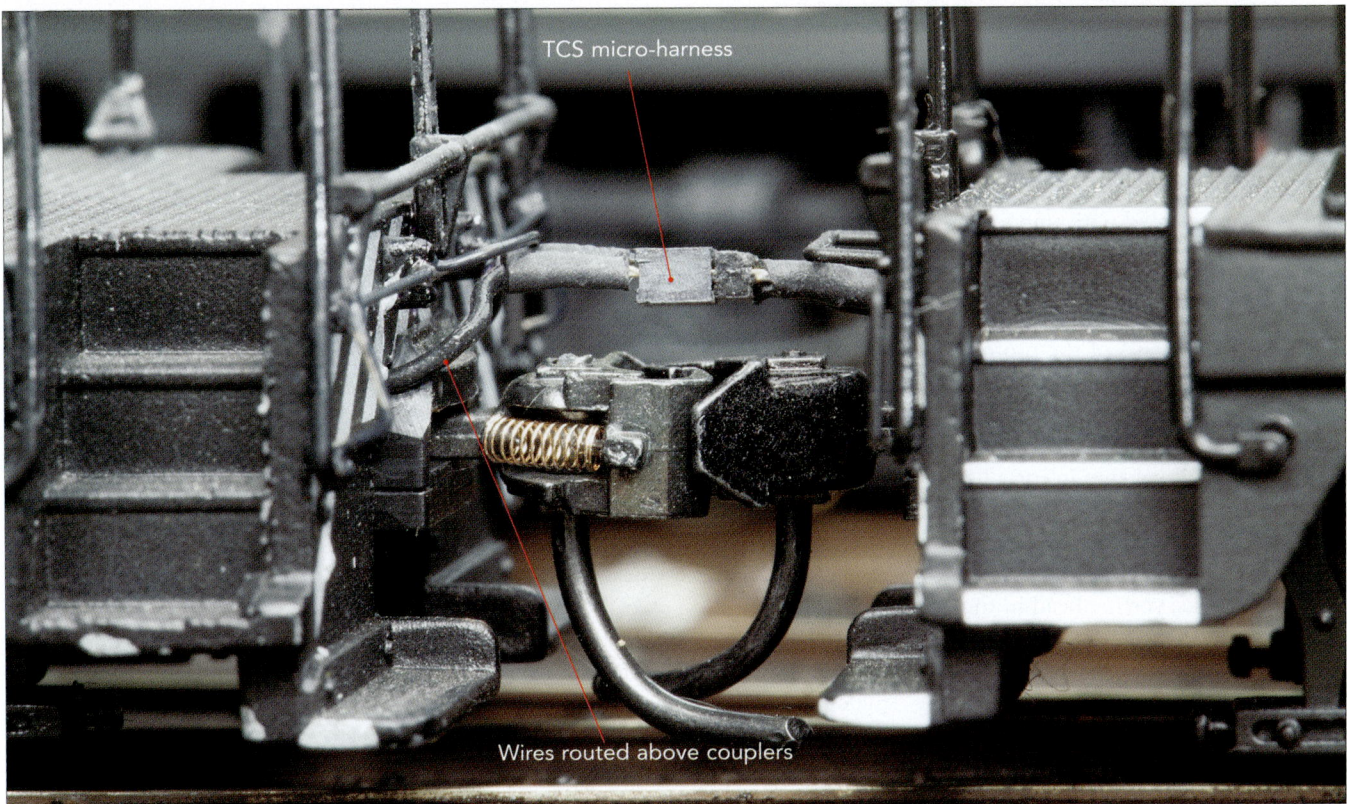

7. A TCS two-wire harness provides the electrical connection for sound between the locomotives.

wires from the front and rear LEDs, respectively, followed by connecting the positive leads to the blue wire.

Next, I connected the purple speaker wires to the solder tabs on the speaker, which I placed over the front truck. I stabilized the speaker by inserting a paper shim between it and the frame. With all the wires connected, I gave both locomotives a test run to make sure they operated in the correct direction and the lights responded properly. I also programmed the addresses to match the numbers on the models.

As you probably noticed, both locomotive models have speakers, but only one has a sound decoder. Since these locomotives are to be operated as a pair in a consist, I powered the speakers in both with the output from just the one ECO-100 decoder. Back on page 60 I mentioned that you can operate two speakers from one decoder, although each will receive only half the wattage.

You can use either series or parallel wiring. In this case I wired them in parallel. This cut the speakers' 8Ω impedance in half. I chose this method so that if I ever separate the locomotives, the sound-decoder equipped model won't have to be rewired, as would be the case if the speakers were wired in series. Also, SoundTraxx recommends the parallel configuration in its literature.

Operating both speakers off one decoder required running a pair of wires between the two locomotives. I used a TCS micro harness for this job so I could quickly and easily separate the locomotives for maintenance and programming.

I soldered the free ends of the harness in the new model to the tabs on the speaker along with the wires from the decoder. In the older model, I attached the two free ends of the harness to the wires I had soldered to the speaker contacts earlier.

The TCS harness has a black and a white wire, so I colored the first inch or so next to the connector black using a felt tip marker to camouflage it a bit. I then shaved off a thin sliver of plastic from the top edges of the coupler mounting boxes and slid the harness wires in alongside them.

This arrangement placed the harness wires and connectors on top of the couplers so they don't get fouled on the coupler trip pins, **7**. With everything installed, I replaced the shells on the chassis and moved to the programming track.

Programming these disparately different decoders so the models operate essentially as a single unit required some pretty fancy maneuvering. Go to page 90 to learn how to speed-match locomotives so they operate at the same speeds throughout the throttle range.

Once that was done I set the decoders up so that the headlights are bright when moving forward and dimmed in reverse. I also used advanced consisting to turn off the rear lights when in a consist, programmed the advanced consist address, and added consist acceleration and deceleration rates.

Visit my website (www.dccguy.com) for more on this project.

3 SoundCar decoder installation

The SoundTraxx SoundCar decoder is made to install in rolling stock to provide prototype sounds including wheels on rails, brakes, couplers, generators, horns, whistles, and light effects. The sound package was selected with unpowered commuter cab coaches in mind, allowing you to install a SoundCar decoder in a cab coach and have all the sounds and lights of the prototype. It also offers prototype sounds for cabooses, which often had whistles for signaling during back-up moves.

The most critical step in installing the SoundCar decoder is providing a way to get power from the track. If you start with a car that has wiper-equipped trucks, you're ahead in the game. However, most cars will require that you install pickups.

I selected wheel and axle wipers from both Richmond Controls (www.richmondcontrols.com) and Streamlined Backshop (www.sbs4dcc.com). These mount on the trucks and ride against the axles or the backs of the wheels. Wires soldered to the wipers conduct power to the decoder in the car. The difference here is important since axle wipers only provide pickup from two wheels per truck, whereas wheel wipers provide pickup from four wheels on each truck, giving more reliable electrical pickup.

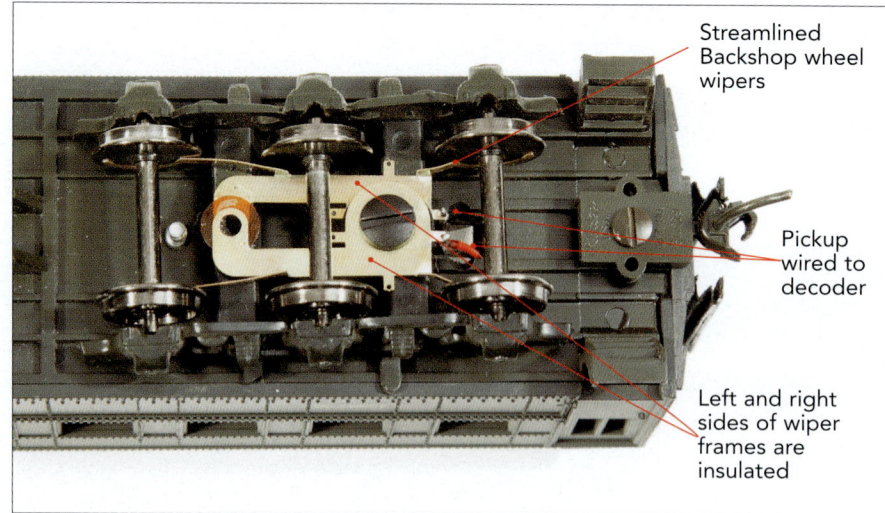

1. Streamlined Backshop's wheel wipers provide excellent power pickup from four wheels on each truck.

I installed decoders in a passenger coach, a boxcar, and a caboose. For the coach's six-wheel trucks I used Streamlined Backshop's wheel wipers, **1**. These come as flat phosphor bronze parts requiring only a couple small folds of the metal wipers before they are installed on the trucks. I soldered wires to the wipers and ran them into the car through holes drilled in the underside of the coach. The coach came with metal wheels on only one side, so I replaced them with all-metal insulated wheelsets, and arranged them to pick up power from both rails.

For the caboose and boxcar, I used Richmond Controls and Streamlined Backshop axle wipers; both slide on over the truck mounting screw and can be bent to engage the axles, **2**. Make sure the two trucks are installed so the insulated wheels are on opposite sides to guarantee pickup from both tracks. The Richmond Controls wipers come with a metal screw that conducts current into the interior of the car, where it can be attached to decoder wires.

After connecting the wires, I attached the SoundCar decoders to the roofs of each car using double-sided foam tape. Be sure to install the flat

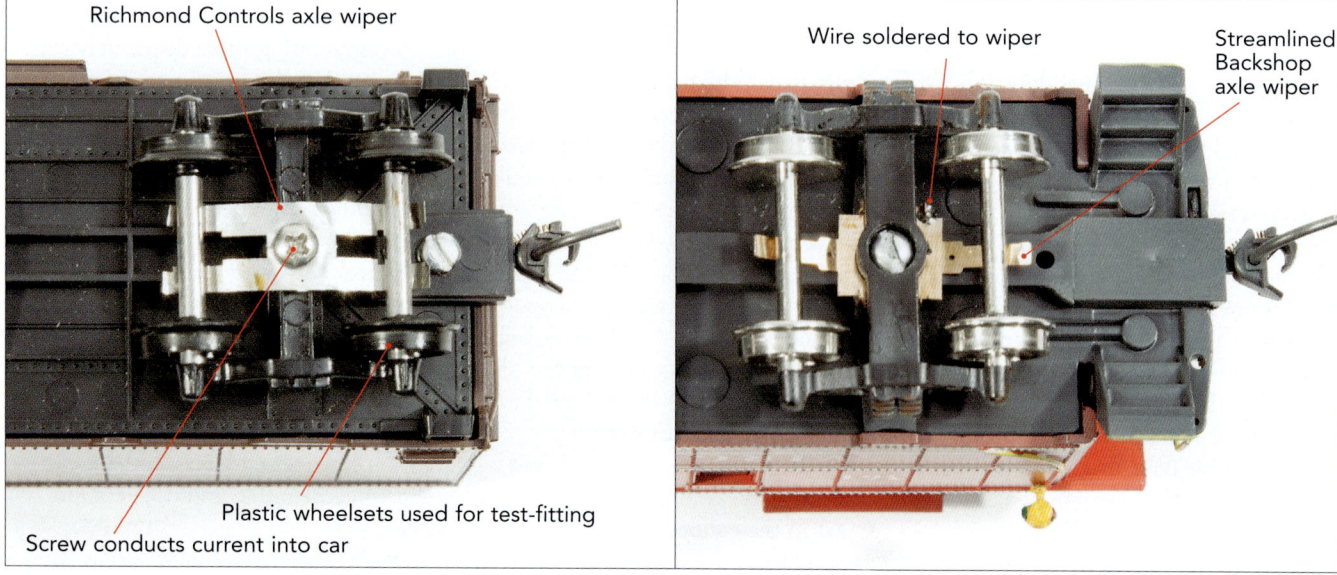

2. Axle wipers like these from Richmond Controls on the left, and Streamlined Backshop on the right, offer another option for power pickup, but may require a SoundTraxx CurrentKeeper for uninterrupted sound.

side of the decoder toward the roof, 3. I drilled holes in the floor of each car in a circular pattern, then mounted the speaker over the holes. I installed a SoundTraxx CurrentKeeper in the caboose to test its performance as well as that of the axle wipers.

With everything installed, I headed for the tracks. To spread the sound out along the length of the consist, I placed the cars in a train with three cars between them. Before pulling out of the yard, I synchronized the sounds of the cars with the movement of the locomotive.

For this, I would normally need to program the locomotive or consist address into each decoder. However, SoundTraxx has made it easy with Intelligent Consisting. All I needed to do was select the locomotive address on my throttle, pass a magnet over the top of the cars, and then press F8 on the throttle four times. This programs the address of that locomotive into all the SoundCars' memories. Waving the magnet over the cars again will clear the memories.

Intelligent Consisting makes the SoundCar respond the same as if I had created an advanced consist. Using Configuration Variables (CVs) 21, 22, and 117, I customized the decoders to control which sounds are available for each type of car, and which work within an advanced consist. I also changed the caboose whistle to a Hancock air whistle by setting CV115 to 2.

On the subject of programming, let me mention that if you use a throttle with buttons for only eight functions, you can swap functions 5-8 with 9-12 by changing CV 30 to 4. This puts sounds like uncoupling, brakes, and coupler clank on buttons 6-8. The procedures for all these changes are covered in detail in the SoundCar Users Guide (www.soundtraxx.com).

With the consist established, the sounds from the cars are synchronized with the speed and direction of the locomotive. As the train began to move, I heard wheels clickety-clacking over rail gaps, flat spots tapping, and flanges squealing. I was very impressed

3. SoundCar decoders should be mounted to the inside roof of the car so that a magnet can reliably trigger intelligent consisting.

with the clarity and quality of the sounds. With the cars separated by several others it gave a stereo effect that seemed to amplify the sounds. However, the sound from three cars was not overwhelming, as I had feared.

I was able to add prototype sounds to switching maneuvers as I hit the brakes and heard the couplers open followed by the sound of the air lines separating. Connecting to a car, I could hear the brakes engage and the couplers clank. Of course, each sound required pressing a different function button, but my fingers quickly learned the steps.

As I expected, power pickup was critical. In the caboose with the CurrentKeeper and axle wipers, the sound effects were bulletproof. In the boxcar without the CurrentKeeper, the sound tended to drop out when going over frogs and other gaps, but even that was not objectionable, since most of the sounds are intermittent anyway. That would be a different matter if lights were installed.

However, with the Streamlined Backshop wheel wipers on the passenger car trucks, there never was a power interruption, even without a CurrentKeeper. Having four-wheel pickup on both trucks resulted in sound almost as reliable as with a CurrentKeeper.

There are a few limitations inherent to DCC sound that require special programming and work-arounds. For example, the horn/whistle can only be remapped among F0, F1, and F2. This creates a conflict with locomotive functions, since you don't want your caboose whistle blowing every time the engineer blows his, and vice versa.

The work-around is to turn off the SoundCar whistle sound in advanced consisting, and independently blow it using the decoder address. This works well if you have a throttle with the ability to control more than one address, or a two-man crew, each with a throttle.

I also needed to adjust the momentum and start values of each SoundCar decoder to match the locomotive's so the clickety clack doesn't start before the locomotive begins to move.

There are numerous other options available for customizing the decoders. If you have a lot of locomotives, you may need to standardize their operations and functions so that your SoundCars will be synchronized across the fleet. Installing these in a fleet of cars will take a bit of planning and programming to get everything synchronized the way you want, but it can add a new level of realism to operating sessions.

4 Use DecoderPro to simplify programming

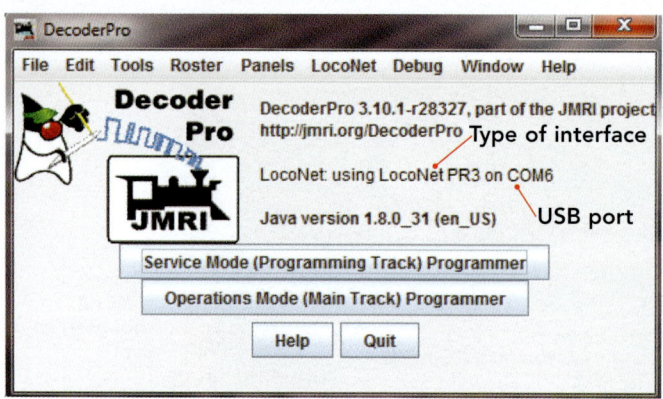

1. When starting DecoderPro, wait for this main pane to open, then either set preferences in the Edit menu or open the Roster.

Using a throttle for programming is fine if all you need to change are a few configuration variables (CVs). However, doing more work than that can get cumbersome when using a throttle. Entering a speed table or doing function mapping can make programming a chore most folks avoid.

For that reason I started using a computer for programming back in 1995 and downloaded DecoderPro as soon as it became available. DecoderPro is one of a package of computer programs called Java Model Railroad Interface (JMRI). It's designed specifically to program decoders. The great thing about JMRI is that it's open-source software, regularly updated by a group of volunteers, and is free from the JMRI website (www.jmri.sourceforge.net).

These folks regularly add new features and updates, and when a new decoder is released, they create a new decoder definition file that users can download. Computers with operating systems as old as Windows 98 are supported; I use an old Pentium running Windows 2000.

You also need a way to connect your computer either to your command station or to a standalone programmer, usually through a serial or USB port. There's an extensive list of supported interfaces on the JMRI website. I tested DecoderPro with a Digitrax PR3, RR-CirKits LocoBuffer, and a Sprog DCC Sprog 3. You can find a more in-depth look at how these three devices work on page 38.

How it works

What are the advantages of DecoderPro? First, all the CVs for the decoder being programmed are right there on the screen, so you don't have to remember them. Plus, instead of showing up as a CV number (for example, "CV3"), it tells you what it is in plain English: "acceleration rate." Using a computer means you can save all the settings for a decoder. That way if your decoder settings get corrupted or reset to factory default, all you have to do is reload the saved settings and reprogram the decoder.

Another benefit is when you have common locomotives and decoders. Let's say you have five Atlas RS-3s all using the same type of decoder. Once you program one, you can simply copy and edit that roster entry, write the settings to the decoder, and then save it in the roster file. Do that three more times and you're done.

After installing JMRI, start DecoderPro by clicking on the screen icon and wait until the main pane opens, **1**. Click the EDIT menu option and pick PREFERENCES—this is where you choose which interface you're using. Select the appropriate choices for your system then hit the SAVE button—the program will ask to restart DecoderPro, and you need to do this before your settings will take effect.

On restart, your interface should now appear in the main pane, **1**; in this case I was using a Digitrax PR3 interface on COM6, which is the USB port the computer will use to communicate with the system.

There are several ways to proceed, but I usually select the ROSTER pane in the main menu, **2**. Note that Direct Byte Mode is selected as the Programming Mode—some decoders do better with this, while others work best with Paged Mode. To get started, hit +NEW LOCO. At this point you can either read the decoder type from the decoder itself or select it from the list.

This is where you may get an error. The JMRI folks told me many

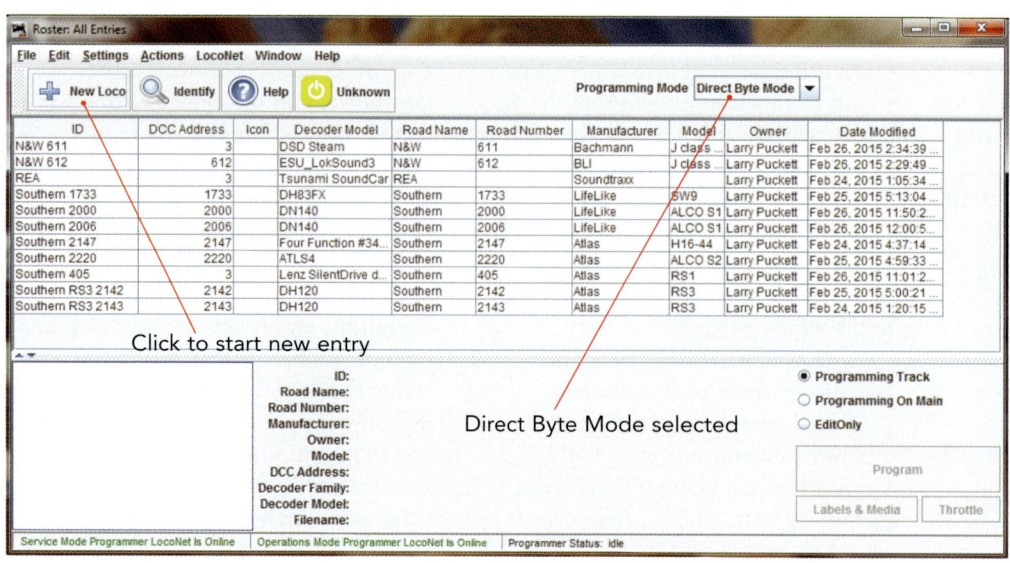

2. The Roster pane provides a listing of the locomotives saved in DecoderPro with brief information on each.

80

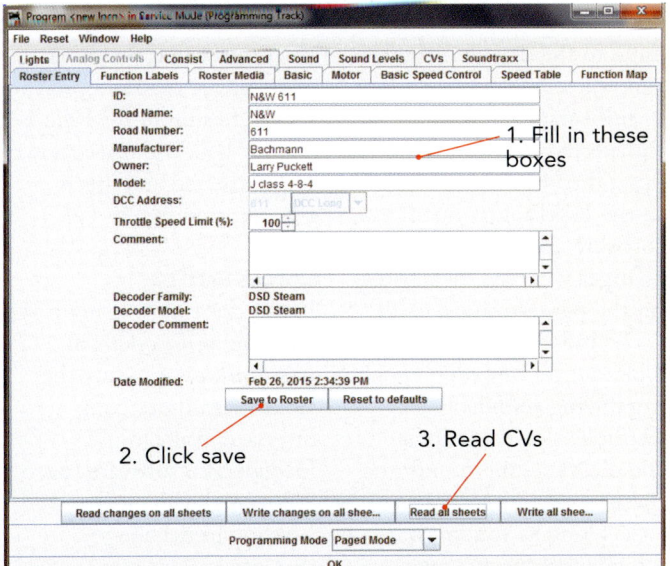

3. Once you open a roster item or create a new one, you can edit the information or move on to other panes by clicking the tabs along the top.

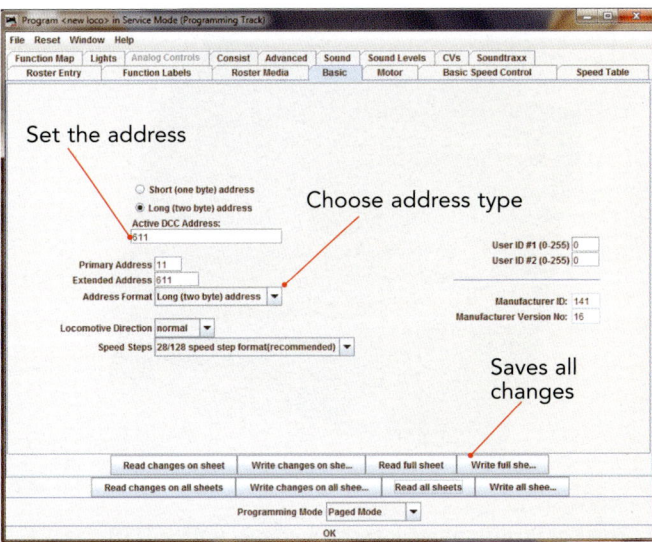

4. The Basic pane allows you to enter or edit the decoder addresses and the information required for CV29, which controls five characteristics.

manufacturers don't assign a unique identification to each decoder, instead using the same software version number for all decoders in a series, which can confuse DecoderPro. Consequently, it's best to just select the correct decoder type from the extensive list and then click on OPEN PROGRAMMER.

At this point you'll see the ROSTER ENTRY pane come up, **3**. Enter all the appropriate information for the locomotive and click SAVE TO ROSTER. Next, hit the READ ALL SHEETS button and wait while the program reads all the CVs. Once this is finished, you can move to any of the other panes by clicking on the appropriate tab at the top.

In the BASIC pane, **4**, you can enter the short and long addresses, set which is active, control the normal direction of the locomotive, and specify the speed steps. In all these panes you have the option of reading and writing just the changes you make or all settings. To be safe, I usually select WRITE FULL SHEET in each pane after making changes.

Some panes only offer a couple selections, whereas some like SPEED TABLE, FUNCTION MAP, and SOUND LEVELS have many.

Speed tables, **5**, allow you to fine-tune speeds. As you can see, DecoderPro provides a simple graphical way to make adjustments. You can use the default curves or move the sliders up or down to adjust each speed step. Once you're done, save the changes and head for the track to see how your new speed table works. It can be helpful to try several different speed tables in the same locomotive just to get a feeling for how they can alter its performance.

If you aren't sure about using DecoderPro, take a few minutes to download it and give it a test drive. You can set it up to operate with a virtual interface simulator in the preferences settings before you decide

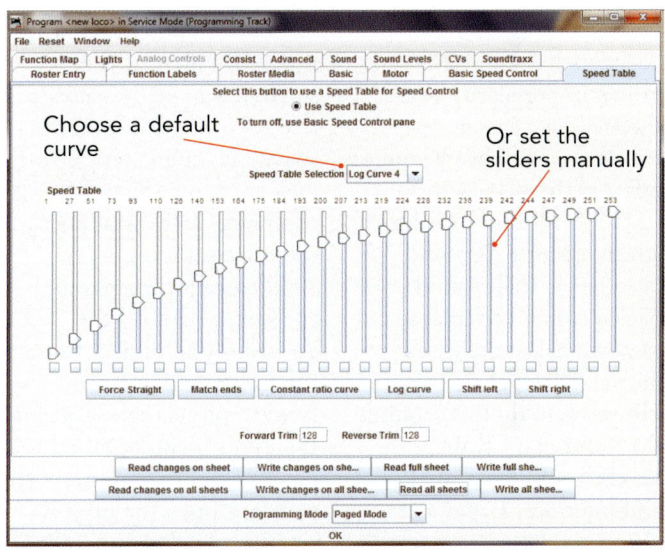

5. The graphical interface simplifies complex tasks such as creating speed tables.

to purchase an interface. And once you get an interface, don't be afraid to experiment with all the CV settings available. If you make a mistake, it's easy to reprogram or reset your decoder and start over.

Because of the low voltage on the programming track you're unlikely to do any physical damage to a decoder, and some interfaces shut down if they detect a short when a locomotive is placed on the programming track. If you run into any problems, there's extensive troubleshooting information on the JMRI website. There's also a JMRI users group on Yahoo Groups (www.groups.yahoo.com).

Tips for programming LokSound decoders

1. The LokProgrammer interface comes with a USB cable, power supply, and options for connecting it to a programming track.

Electronic Solutions Ulm (ESU) LokSound decoders have rapidly grown in popularity over the last few years as several manufacturers have begun offering them as factory-installed options. One big advantage of LokSound decoders is their sound packages can be replaced as new ones are released on the ESU website. However, to do this requires the proprietary ESU LokProgrammer software and interface, **1**, and a Windows-based computer.

If you'll be doing a lot of LokSound decoder updates, the LokProgrammer software and computer interface hardware really is your only option unless you plan on shipping the decoder back to a dealer each time. The interface has solder pads as well as screw terminals for connecting feeder wires to a programming track.

Once you install the LokProgrammer software on your Windows-based computer, plug the interface hardware into a USB port on your computer and install the device drivers. The software can be downloaded free from the ESU website (www.esu.eu/en/start/). You may need to restart your computer after installing the drivers, then start the LokProgrammer software.

Next, download the desired sound project from the LokSound website (www.projects.esu.eu). Using the Files menu option on the LokProgrammer software, open the project you downloaded and let it load into the program.

There are different types of sound projects—Version 4.0 projects can be edited and the sounds modified. With LokSound Select projects you can only change the CV settings. Once the project loads into LokProgrammer, you can edit the various CVs to change the address, momentum settings, and so on, **2**.

After making all your edits in the decoder software, it's a simple matter to select "Write the decoder data" and "Write the sound data" to the decoder. This can take a while, since a lot of information has to be transferred. You can then open the virtual throttle in the LokProgrammer software and test your changes with the locomotive on the programming track.

Sound slots are LokSound's way of organizing sounds. For example, slot 1 usually is the diesel engine sound in a diesel sound project, **3**.

As long as you know which sound is assigned to which slot in the sound project, there's no problem. However, they may be different in different projects. These slot assignments are available on the ESU website along with each sound project. Just save or print a copy and keep it for your records. Also, there's an option in the LokProgrammer software "Tools" menu to print out a bulletin giving the slot assignments.

The function output controls, **4**, allow you to modify how the lights and any other wired functions operate. For example you can set the headlight for Rule 17, and in the locomotive shown set up the red and green classification lights.

DecoderPro

If you don't have many ESU LokSound decoders, or you don't want to buy the LokProgrammer computer interface, you can use DecoderPro software (page 80) to make changes to LokSound decoders.

Let's look at how DecoderPro handles LokSound projects. Because there are few DecoderPro definition files for most current LokSound sound projects, you need to get the CV settings into DecoderPro another way.

One option is to place the locomotive with the installed decoder on the programming track and attempt to read the CVs with DecoderPro. This approach can take a long time, and you will likely end up with a lot of read errors.

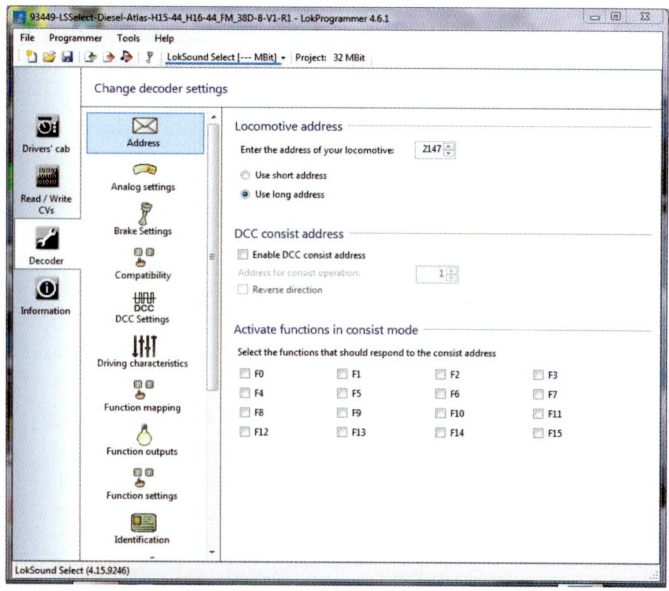

2. The LokProgrammer Decoder screen allows you to edit CV settings such as the Address.

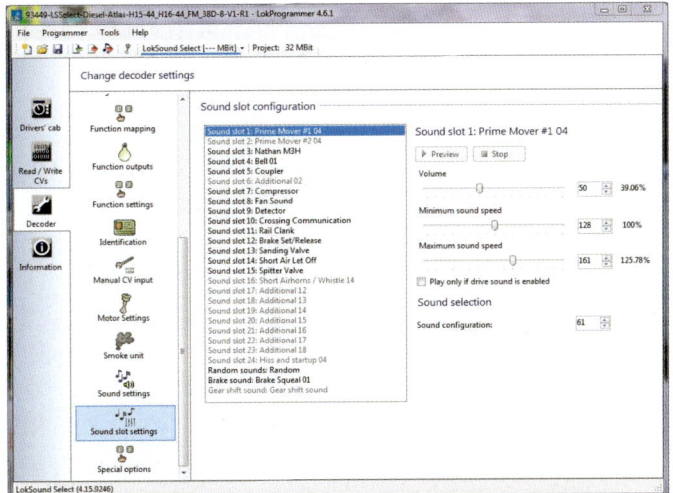

3. On the LokProgrammer SOUND SLOT SETTINGS screen you can change things like volume, minimum and maximum playback speed, and whether the sound plays when the locomotive is moving or stopped.

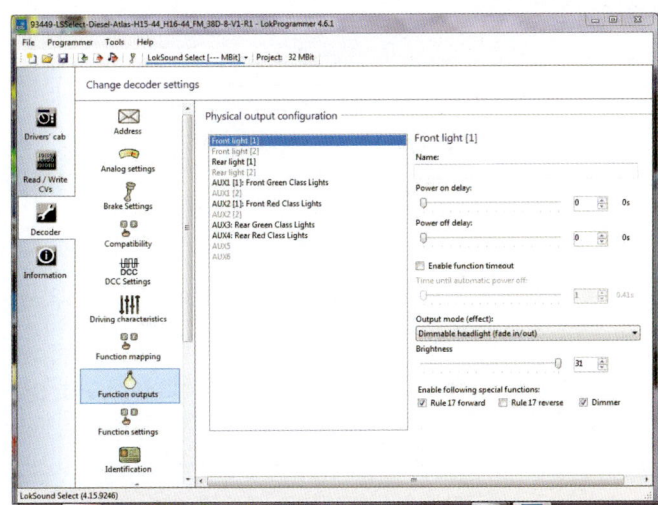

4. The LokProgrammer FUNCTION OUTPUT screen is where you control wired outputs like the lights. For each output you can control the type of light effect, whether it's dimmable, and whether Rule 17 applies.

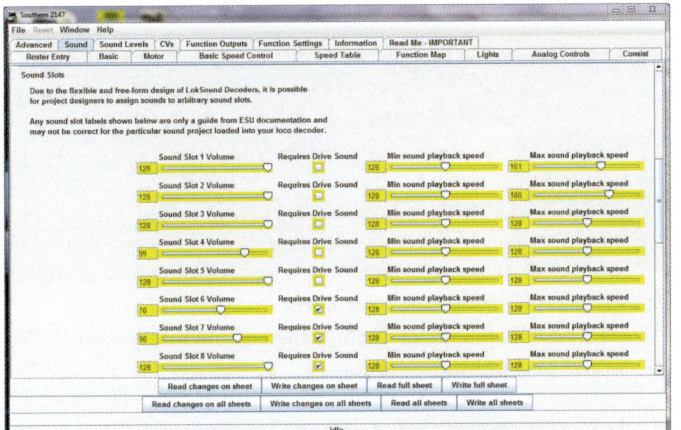

5. In DecoderPro the sound slots are edited in the lower portion of the SOUND pane. It has controls similar to the LokProgrammer sound slot screen.

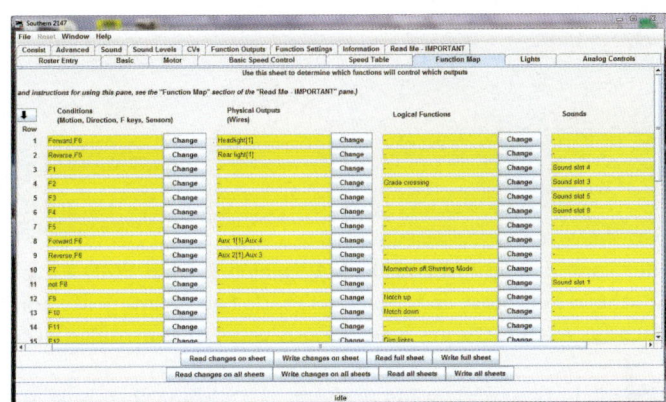

6. The DecoderPro FUNCTION MAP pane allows you to assign functions to specific buttons, designate the characteristics of each, and assign sound slots.

The alternate option is to install the LokProgrammer software and use it without the interface hardware. You don't need the hardware interface to examine and edit sound projects.

As I described above, open the LokProgrammer software and read in the file you downloaded from the ESU website. Now, using the TOOLS option, export a CV list file.

Open DecoderPro, create a new locomotive entry selecting the LokSound Select decoder you have, and open the comprehensive programmer. Using the FILE > IMPORT > LOKPROGRAMMER CV LIST FILE option, read in the CVs from the LokProgrammer CV list file. You will now have all the CV settings loaded into DecoderPro.

Once you get all your CV settings into DecoderPro, it's equally easy to edit them as I've described for other decoders (page 86, for example) and in a series of videos on my website (www.dccguy.com).

The sound slots are edited in the lower half of the SOUND pane, **5**, giving you control over volume, Min and Max playback speed, and a checkbox to set whether the sound plays with or without the locomotive moving.

More advanced controls in the FUNCTION MAPPING pane, **6**, allow you to assign sound slots to specific function buttons, as well as physical and logical output controls.

If you don't have DecoderPro, you can still make all the edits using the LokProgrammer software. Then, using the TOOLS > SHOW CHANGED CVs tab, save these changes to the computer's clipboard, and enter them in the decoder manually using your DCC system's programming throttle.

Page 52 shows the installation of a LokSound decoder in an HO scale Atlas H16-44.

4 Getting the most from automatic functions

Hurricane damage forced these Virginian & Ohio EMD F3 and Alco FA diesels to detour over Larry Puckett's Piedmont Southern rails. Just as full-size railroads had to integrate control systems on locomotives from different manufacturers, DCC decoder manufacturers have various ways of controlling automatic sound and lighting effects.

I explain how remapping decoder functions (page 86) makes it easier for operators to access the growing number of user-triggered functions available with decoders from different manufacturers. However, there are also a growing number of automated functions and features that can be activated without operators having to touch a button, or which occur as a side effect of another action. Let's take a look at a few of these as offered in some of the most common sound decoders.

Brakes

One feature common to many sound decoders is brake squeal. At one time you had to hit a function button each time you wanted the brakes to squeal. In most current decoders the brake squeal is activated when some threshold is reached, usually due to a change in throttle setting. However, there can be some interesting interactions.

For some time I tried to make the brakes squeal on my ESU LokSound Select-equipped Atlas Alco S-2 switcher. I tried all kinds of different combinations of on-and-off thresholds using configuration variables (CVs) 64 and 65, to no avail. Then someone on the LokSound Yahoo Group mentioned that you need to crank up the deceleration momentum value in CV4. I had set the value to 12, and that was too low a threshold for an audible squeal. Setting CV4 to a value of 30 brought the brakes to life and a value of 80 really drew them out when I reduced the throttle setting.

SoundTraxx Tsunami and Train Control Systems WOWSound decoders have their own idiosyncrasies with brakes. With SoundTraxx you have to turn on automatic brake squeal by setting CV198 to 4, and then adjust the sensitivity setting in CV196 (default is 3). The brakes then squeal when function 11 braking is used or when there's a rapid reduction in the throttle setting. WOWSound uses a minimum speed step setting for the point at which brake squeal starts. This is entered using indexed CVs 201 to 204.

If you're not happy with the default setting of 15, I suggest you go to the programming section of the WOWSound website for the exact values if you don't use Java Model Railroad Interface's (JMRI) DecoderPro. (See page 80 for information on DecoderPro).

The brakes on WOWSound decoders only squeal when brakes are

applied using F7. The brake sounds are changed at random, so they're a little different every time.

Throttle notches

Another neat sound feature many folks don't use or know about is the ability to adjust the automatic notch rate with some decoders. This feature allows you to change how quickly the locomotive engine RPM sounds ramp up as you advance the throttle.

With SoundTraxx Tsunami2 and Econami decoders you can use CV114 to set the number of speed steps between the notches at anywhere from 1 to 15 steps per notch. Older Tsunami decoders use CV116. This makes a big difference in the sound you get from your decoders as you advance the throttle.

The SoundTraxx approach forces the increase in locomotive RPMs to a constant rate of change. Set it to 7 and you max out at about speed step 50 (assuming you are using 128 speed steps). If you operate your locomotives at prototypically slow speeds, that's fine. At speeds above step 50 there will be no more increase in RPM.

LokSound Select decoders basically spread the notches out evenly over the lower half of the speed table, and the rate can't be changed.

With WOWSound decoders you can set the number of steps between each notch independently. That way you can bunch up the RPM increase in the lower speed steps, then spread it out more in the higher speed steps as the locomotive picks up speed.

As with the brake squeal setting, the notch steps are entered using indexed CVs 201 to 204. The programming section of the WOWSound website has the exact values. I find the notch step feature especially useful with yard switchers that don't often get out of the lower speed step range, but may have to notch up a bit to shove a long cut of cars around.

Also keep in mind that all these decoders offer some form of manual notching, which allows the engineer to control the point at which RPMs increase. Manual notching usually works by controlling the speed with the throttle and using function buttons to change the RPMs.

Headlights and reversing

Now let's talk about Rule 17. As I explain on page 86, this actually covers an array of rules related to the use of headlights. In most cases, decoders offer automatic headlight dimming whenever the locomotive stops.

All three of these manufacturers' decoders take this a step further, allowing you to set up the headlights so both the front and rear lights can remain on in both directions, with dimming when running in the opposite direction.

WOWSound decoders are probably the easiest of the three to program for Rule 17. I just write a value of 56 to CV61 and then program CVs 49 and 50 to a value of 40. SoundTraxx is more complex, since the CVs involved also program a large array of lighting features. Setting CVs 49 and 50 to 145, and on Tsunami2 and Econami decoders CVs 57 and 58 to 253, did the trick in my case.

LokSound decoders require changing several function CVs after setting index CV31 to 16 and CV32 to 2. Once these are set, program CV257 to 16, CV266 to 1, CV268 to 0, CV273 to 16, and CV282 to 2. If you have problems, try resetting the index CVs before programming each of the function CVs.

Finally, here's one that may drive you and your engineers crazy. It's possible to set features to occur when the locomotive reverses direction. For example, you could set it to sound the horn and ring the bell automatically each time the locomotive reverses direction. But this would probably get old fast, and most railroads didn't require this when doing repeated switching maneuvers.

Other automatic features can be programmed to activate when the grade crossing whistle or horn is activated. One typical use would be to have the ditch lights flash and the bell ring when the horn is blown for a crossing.

Most of what I've discussed isn't a concern when locomotives are operated individually. However, consists can complicate things. Some of the differences in the ways decoders implement various functions are negligible, but others may take some planning, and a few make it difficult to operate mixed consists.

For example, a WOWSound decoder's F7 braking function and LokSound's Full Throttle brake feature would fight one another. Also, these features wouldn't work well together in universal consisting, since the brakes would only activate on the lead locomotive.

I know of some folks who've gone whole-hog and decided to standardize their locomotive fleet using one brand of decoder. But replacing all your decoders would be a prohibitively expensive (and time-consuming) option for most modelers.

With a little advanced planning when installing decoders, you could easily use the same types of decoders in locomotive models you plan to operate in consists. Similar functions can be remapped to the same throttle buttons.

Using advanced consisting, as I describe on page 88, allows you to control which functions respond to throttle commands when operated in a consist. More importantly, there are some differences in functions that don't matter in a consist. For example, you don't have to worry about headlights, horns, and bells in the center locomotives, since they would be off anyway.

Programming all these features to work automatically the way you want them to can get a bit complicated. In some cases it can be difficult to find out how various interactions work, or the function may not be fully documented in the manuals. Consequently, I've found DecoderPro to be the best reference for figuring this out. In most cases the sliders and selection boxes are grouped according to function, making it easier to set up complex features.

For more on programming these and other decoders, visit my website at www.dccguy.com.

4 Remapping functions for consistent operation

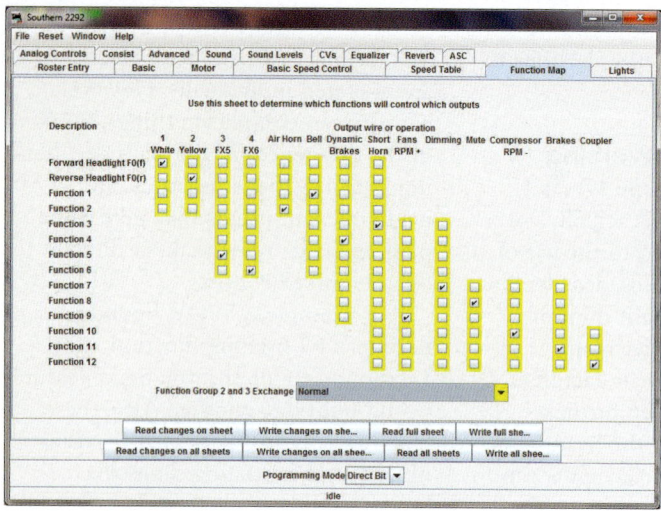

1. The old SoundTraxx function mapping approach for a Tsunami decoder, shown here in DecoderPro, displays the limited range of function buttons available for various light and sound effects.

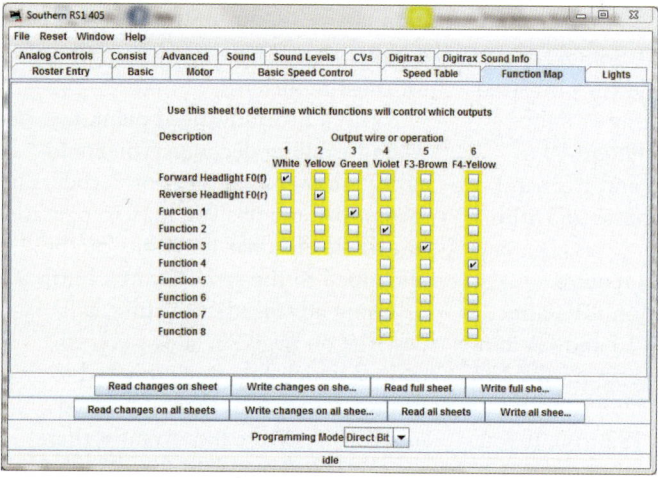

2. The Digitrax sound decoder function mapping pane in DecoderPro is similar to the one shown in 1, but note there are no sound selections available. Function button assignments for their sound decoders are programmed into the sound project files, which must be edited to change button assignments.

With the current generation of DCC sound decoders, there are so many user-controlled functions it can be difficult to remember which buttons to push on your throttle. Unfortunately there's little consistency from one manufacturer to the next when it comes to which functions and sounds are assigned to throttle buttons above function F2.

For example, SoundTraxx uses F11 for brakes, but TCS uses F7. During an operating session you may get used to using one button for the brake on your first run, then have to re-learn which button to push when you head out with a second train.

Fortunately, DCC decoders support function (re)mapping to varying degrees. This feature allows you to assign sound, light, and control functions to specific buttons on your throttle.

For example, with the SoundTraxx Tsunami decoder, you could move the brake function from F11 to F7 so it would be the same as your TCS WOWSound decoders. You can also combine multiple functions so they're activated by the same button.

An example of this would be to configure the horn button so it also activates ditch lights or the bell each time it's blown. One limitation of this is that with the SoundTraxx Tsunami approach in DecoderPro, only certain functions can be assigned to a small range of buttons. DecoderPro is a tool in the Java Model Railroad Interface (JMRI) used to program decoders (www.jmri.sourceforge.net; see page 80). As you can see in **1** and **2**, the headlights can only be assigned to F0 through F2 or F0 through F3 depending on the decoder.

Over the years, this function-mapping feature has become more complex as the number of functions has increased. Currently, the NMRA recognizes functions F0 to F28, and most manufacturers provide either detailed manuals or support for them on their websites.

Both Digitrax (www.digitrax.com/support/cv/calculators) and TCS (www.tcsdcc.com) provide programming tools that lead you through the selection process, and then provide the values to be written into the required Configuration Variables (CVs). TCS also offers Audio Assist on its WOWSound decoders. This feature gives voice prompts and feedback to help program sound effects.

When it comes to remapping these functions, I usually turn to DecoderPro, even when using the "legacy" SoundTraxx approach. With DecoderPro you simply click on the check box where the effect and function button intersect, **1**.

Digitrax has taken a different approach with its sound decoders, **2**, making sound effect assignments part of the sound project itself. The downside to this approach is you have to edit the sound projects in order to make assignment changes.

With the new SoundTraxx Flex-Map approach in DecoderPro, introduced in the firm's Econami and Tsunami2 decoders, each effect has a drop-down menu allowing you to assign it to one of the 29 (F0 to F28) function buttons, **3**. If you choose to program the effect using a throttle, you first have to program CV31 to a value of 1 to access the indexed CVs, and then program the desired function button number into the corresponding indexed CV. LokSound also uses indexed CVs, and I recommend using either its own LokProgrammer or DecoderPro.

One thing about DecoderPro is that the computer code for various function-mapping panes appears to be written by different people, and as a result the panes may look a bit different. Some have the effects listed in order

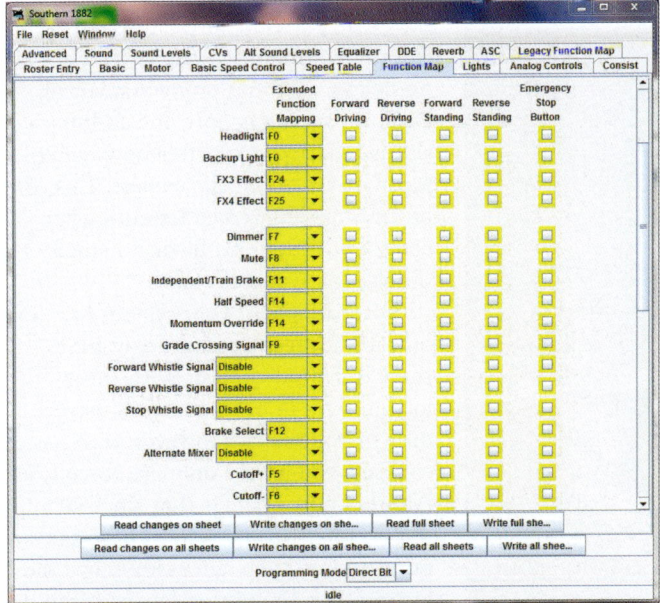

3. SoundTraxx's new Flex-Map approach, a feature of the new Econami and Tsunami2 decoders, allows you to assign any light, sound, or logical effect to any of the available function buttons. It also allows automatic controls keyed to direction and movement.

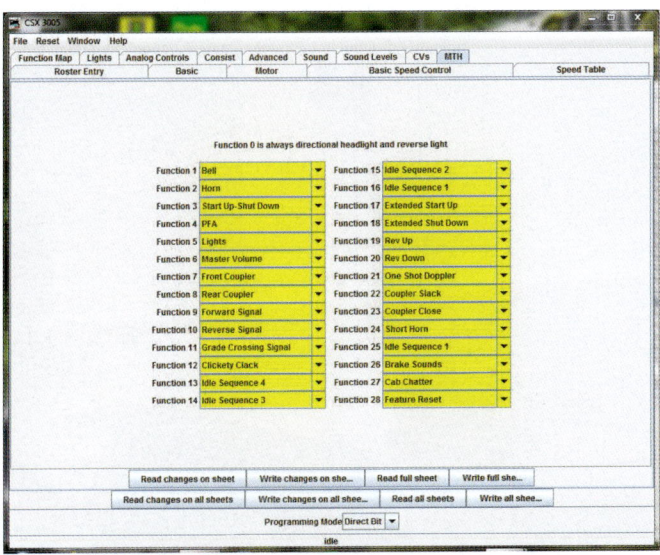

4. The DecoderPro function mapping pane for MTH decoders presents the function buttons with a drop-down menu of effects next to each. The various function mapping panes look different to take advantage of each decoder's unique features.

and drop-down selection boxes or check boxes for the button numbers, 3. Others take the opposite approach, using function number lists with drop-down boxes associated with the effects, 4.

The LokSound pane uses a multiple selection approach, with even more flexibility, 5. While this may seem confusing at first, you'll find it's often done to take advantage of unique features of the decoders.

Methods

With all the sounds, lights, and controls available, it can be difficult to keep them all sorted out. Therefore, having a logical way to pick and choose among them is important, requiring a systematic approach. My method is to make a table or spreadsheet of the functions offered by the various brands of decoders I own.

Once I have my spreadsheet of available functions, I whittle those down to a common group I think my operators will actually use. Finally, I decide which of those functions will be assigned to the same buttons on my throttles for all my decoders. This can be complicated by the fact I have a collection of throttles with anywhere from 13 to 29 function buttons, if you count the ones accessed using a "shift" button.

Because it can be difficult for an operator to hold a throttle and car cards while pushing a shift and function button, I've further narrowed the selections to just those directly accessible on utility throttles. In my case, that's F0-F6. So in this process I've gone from more than 29 possible sound, light, and control functions to only seven.

This may seem like a major waste of all those neat sound and light effects you paid for. However, there are

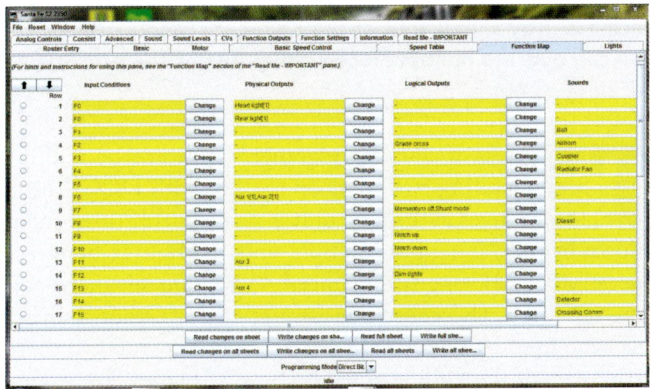

5. LokSound decoders offer yet another DecoderPro function mapping pane allowing you to change physical, logical, and sound assignments for each button on the throttle.

other ways to take advantage of them without pushing buttons. Many decoders offer varying degrees of automatic sound and light effects. For example, with some you can set your brakes to squeal when locomotive speed decreases at a preset rate. A sensitivity setting allows you to fine tune this so the brakes stop squealing before the locomotive stops.

Bells and lights can be set to activate when a whistle or horn signal is blown. Effects may also be triggered when locomotives change directions or stop. In many cases, you're only limited by your ingenuity and imagination.

4 Basic, universal, and advanced consisting

SD40-2 no. 1027 leads a unit coal train through Williams Bay on the Milwaukee, Racine & Troy club layout at the Kalmbach Media offices. Using DCC consisting you can operate multiple locomotives just as the prototype does.
Bill Zuback photos

The process of coupling together and operating two or more locomotives as a single unit is known by several names. Multiple-unit operation or MU-ing (pronounced "em-you-ing"), also called consisting, has been used by railroads since the early years of the 20th century, first with electric locomotives and then with diesels.

For prototype diesels, once a group of locomotives is coupled together and set up for multiple-unit operation, all the control operations in the lead locomotive are simultaneously sent to the other locomotives so they operate as a single unit. To accomplish this, both air and electrical lines are connected between locomotives to control the throttle, brakes, lights, dynamic brakes, and sanders. This capability gave diesels a big advantage over steam, as double-headed steam locomotives always required individual crews.

When it comes to our model locomotives, we can easily imitate prototype multiple-unit operation with DCC. This is accomplished using one of three methods referred to in Digitrax literature as basic, universal, and advanced consisting.

Other sources and manufacturers may refer to these methods by different names. For example, I've seen universal consisting referred to as basic, brute force, and old-style consisting. To avoid confusion, I'll describe what each does and you can use whatever name suits you.

Using these three methods we can control the speed, direction, lights, whistles, bells, and other functions that we normally use with a single locomotive. Let's take a look at each of these methods and their pros and cons.

Basic consisting

Basic, or address, consisting is the simplest form of multiple unit control. All that's required is setting the decoder address in each locomotive to the same value. Since all locomotives in a basic consist have the same address, they'll all respond simultaneously to throttle commands—there will be no delays. Another point in favor of basic consisting is only one address slot is required no matter how many locomotives are in the consist. This can be an important consideration when using DCC systems having a small number of available slots.

For example, if your system has only 10 address slots, and you have three universal consists with three locomotives in each, that will use up nine of your slots. However, with basic consists, you would only use three slots. Another advantage is that since all the decoders in the consist have the same address, you can easily move the locomotives to another layout without having to re-create the consist.

There are some limitations to basic consisting. All the locomotives in a consist must be facing the same direction, unless you change configuration variable (CV) 29 so a locomotive you want to run backward is set to "reverse" for normal direction of travel. This will allow the lights to automatically reverse in sync with the consist, although in prototype practice, the lights on a trailing unit wouldn't be on.

The real issue in basic consisting is with sound effects, since you can't control the functions in each locomotive separately. If you blow the horn, ring the bell, or activate other sounds, all the locomotives will respond. This isn't very realistic, and can create quite the cacophony especially if you have three or more locomotives running together.

Universal consisting

Universal consisting is probably the most popular. This method, also known as command-station-assisted consisting, works by having the command station keep track of each locomotive in the consist. To do this, it uses a memory slot for each locomotive address.

All the DCC systems I've used have some way of initially setting the direction of the locomotive so no special reprogramming or rewiring is required. The first locomotive entered when the consist is created becomes the top or lead locomotive, and the command station keeps track as additional locomotives are added.

However, the address of the first locomotive remains as the consist address and all commands are sent to it first. This is where time lags can creep in. Because commands are sent sequentially to each locomotive in the consist, you can visibly see response delays. To test this, I lined up five locomotives in a consist, turned on the headlights and reversed direction—the headlights in each locomotive came on in the order in which they had been added to the consist, one after the other, not simultaneously.

Why is this an issue? The same thing happens when you alter throttle settings, creating visible delays in how fast the locomotives respond. Since the command station has to keep track of all the locomotives being operated, these delays get longer as the number of locomotives being operated in the consist increases.

Another downside to universal consisting is that if you move the locomotives to another layout, you'll have to re-create the consist there. That's because the command station has all the information about the consist in its memory. Also, if you have to reset the command station or its internal battery goes dead, you'll have to re-enter your consists.

On the positive side, the delay usually doesn't cause problems on most home and even most club layouts where few locomotives are actually running at the same time. Also, even though a locomotive is part of a consist, you can still select its address and control individual functions—for example headlights can be turned on and off and individual sound effects like horns and bells can be triggered.

Advanced consisting

Finally, there's advanced consisting, which offers many advantages over the other methods. With this method the same consist address is entered into CV19 in each locomotive's decoder.

(Some older decoders don't support CV19.)

This means the consist is independent of the command station and can be moved to another layout and operated without any reprogramming. Also, because each locomotive decoder is responding to the same address, there are no delays, and each consist only uses one address slot.

Using CVs 21 and 22, you can control how each of the first eight functions DCC responds to commands when in a consist. For example, you can turn off lights and sounds in all the middle or trailing units and only allow the first unit to respond. Some decoders now offer other advanced consisting functionality. The biggest downside to advanced consisting I know of is you're limited to consist addresses 1 to 127.

Setting up consists can require some planning and programming. With advanced consists in particular you need to decide how you want the functions on individual locomotives to respond when in a consist and set CVs 21 and 22 accordingly.

Of course this may all change if you switch a mid-consist locomotive to a lead position. Clearly, once a consist is set up there are disincentives to altering it in the future.

Depending on the capabilities of your DCC system and throttles, you may find it easier to use DecoderPro to program your decoders for advanced consists through the Java Model Railroad Interface (JMRI) on your computer. There's also a tool for setting up consists.

Another factor that needs to be dealt with for all types of consisting is ensuring that the locomotives to be consisted all operate at approximately the same speeds. Speed matching is the process of programming the decoders in the locomotives so their speeds match at all points throughout the throttle range. Failure to do this may result in poor performance. You can learn about speed matching on the following pages.

Consist styles pros and cons

Basic (address) consisting
PROS:
Simple
Instantaneous response
Only one address slot needed
Easy to move to other layouts
CONS:
Locomotives must face same direction
Sound effects triggered on all locomotives simultaneously

Universal consisting
PROS:
Direction of locomotives can be set individually
Sound effects can be triggered individually
CONS:
Time lag between units as commands are sent
Consist can't be used on another layout because information is stored in command station

Advanced consisting
PROS:
Consist address stored in each locomotive's decoder, so consist is portable
Direction, lighting, sound effects are set for each locomotive
No time lag in response to commands
CONS:
Limited to consist addresses of 1 to 127
Changing consist requires multiple steps
Requires decoders that support CV19, which is absent from some old decoders

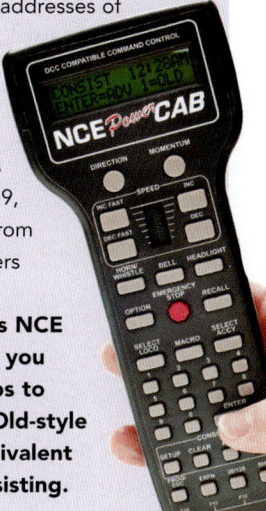

The menu on this NCE PowerCab takes you through the steps to build a consist. Old-style consisting is equivalent to universal consisting.

4 Speed matching for DCC consists

1. A pair of Southern SW1 switchers passes through the Model Railroad Technologies Accutrack II speedometer at 4.1 scale mph. The Accutrack II makes speed matching quick, easy, and accurate, and doesn't require permanent installation on your layout.

On page 74, I showed how I installed decoders in a pair of Walthers SW1s to be operated as a two-unit consist. Any time two or more locomotives are operated in a consist, you should first must make sure they run at about the same speed across the throttle range. Speed matching is the process of adjusting the decoder output to achieve this goal, **1**.

Depending on the type and age of the locomotives, you may not need to make any adjustments. For example, locomotives from the same manufacturer may run at essentially the same speeds, whereas models from different release dates or from different manufacturers may need major adjustments.

Locomotive models may also run at noticeably different speeds when operated in different directions. This can be corrected using forward and reverse trim decoder settings. Some decoders only support using forward and reverse trim with 28-step speed curves.

Back-electromotive-force control (back-EMF) is another potential complication. It allows a decoder to monitor and control speed based on load, and can contribute to smooth running. Any time two or more locomotives are operated in a consist with back-EMF on, they may end up fighting other locomotives in the consist that don't have back-EMF. Dealing with this may require turning back-EMF off or reducing its intensity.

When speed matching I generally recommend using a three-step speed curve approach. This is fast and easy and works with most locomotives and decoders. I use DecoderPro in the Java Model Railroad Interface (JMRI), a free download at www.jmri.sourceforge.net.

You can also use a DCC throttle to enter values into the configuration variables (CVs). Three-step speed curves use configuration variables 2, 5, and 6 to set the starting speed, top speed, and mid-point speed, respectively, **2**. Some older decoders don't support CVs 5 and 6, in which case you will need to use a 28-step speed curve, **3**, and some don't support the use of CV2 in some cases.

Programming

I usually do three-step speed curve entries with ops mode programming using either a throttle or DecoderPro, **2**. This allows me to quickly optimize the settings for CVs 2, 5, and 6. For more complex programming, I generally use DecoderPro. This program organizes tasks by function, making it easier to figure out how to do more complex tasks. You can also use a 28-step speed curve to mimic a three-step speed curve by setting steps 1, 14, and 28, then straight-lining the intermediate steps, **3**.

Proceed in consistent steps and keep notes as you make changes. This will help you keep track of what works and what doesn't. Here's my process:

1. Turn off back-EMF if possible, or reduce the intensity. This may not be desirable with some decoders because of complex speed control algorithms and interactions.

2. Set momentum (controlled by

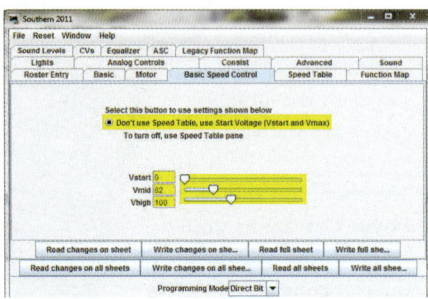

2. The Basic Speed Control pane in DecoderPro allows you to enter values for CVs 2, 5, and 6 either by keying them in or with the sliders.

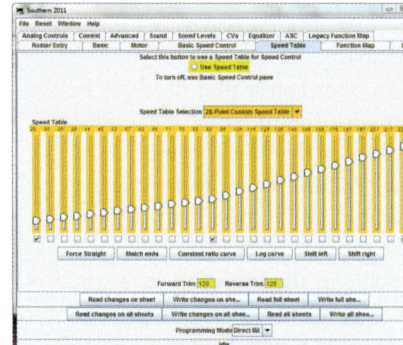

3. Using the Speed Table pane, you can create a three-step curve by designating speeds for steps 1, 14 and 28 and straight-lining the steps between.

CVs 3 and 4) to "0."

3. Set all locomotives to move at the same starting speed using CV2 (starting voltage).

4. Set all locomotives to run at the same top speed using CV5.

5. Set all locomotives to run at the same midpoint speed using CV6. With DCC throttles that display speed as a percentage (e.g. Digitrax DT400/402/500) I use 1 for the minimum, 50 for the midpoint, and 99 for the maximum.

6. Make adjustments for differences in forward and reverse speed using the forward and reverse trim CVs. You have to enable speed curves using CV 29 in order to do this.

7. Once you have all the locomotives running at about the same speeds, then you can begin to restore features like back-EMF and momentum.

Judging locomotive speed can be tricky if done by eye. For accuracy and reproducibility, I use an Accutrack II scale speedometer from Model Railroad Technologies, **1**, available from Streamlined Backshop (www.sbs4dcc.com) and others. It's entirely stand-alone and portable, has built-in calibration for HO (1:87.1), OO (1:76), and N (1:160), and can display speeds in miles per hour or kilometers per hour. The most important thing about the Accutrack II is that it simply sits across your existing track—there are no detectors to install. The locomotive passes between two detector beams within its enclosure, and a microprocessor in the unit calculates the time this took and converts it to scale speed.

Because my old SW1 model with the TCS decoder ran much slower than the new one with the SoundTraxx Econami decoder, I ended up increasing the starting voltage of the old model and then dropping the maximum and midpoint voltages of the new model. Fortunately the back-EMF settings of the decoders were not an issue.

With the switchers operating back-to-back in a consist, I turned the rear lights off. The trick to separating them is simple enough—just remap the yellow wire powering the rear headlight to another function button (see page 86 for more on function mapping). With DecoderPro you can do this with a couple mouse clicks. Another option is to disconnect the yellow wire from the LED or bulb.

Typically in yards, the forward light on a switcher was on full and the rear light was dimmed. To accomplish this with a pair of locomotives I set the headlights in each locomotive for Rule 17 auto dimming. The TCS and SoundTraxx decoders will then automatically dim the headlight when the locomotives stop or reverse direction.

For those of you without DecoderPro, set CV61 to 49 in the TCS decoder and to 50 in the SoundTraxx Econami decoder. Also set CV49 to 40 and CV50 to 0 in the TCS decoder and 145 and 0 respectively in the SoundTraxx decoder for control of the forward and rear lights. In the SoundTraxx Econami decoder, CV57 and CV58 should also be set to 1 so that the headlight remains on in both forward and reverse directions.

Now let's get these locomotives ready for an advanced consist. I designated SW1 2011 as the lead unit and programmed a value of 11 into CV19, the advanced consist address. Because the normal direction of travel for unit 2008 was to be reverse, a value of 139 was programmed into CV19 on its decoder. Adding 128 to the consist address sets the locomotive for reverse operation.

Next came the settings for advanced consist function control in CVs 21 and 22. With the older unit I only needed to control the headlight, so I set CV22 to 1, allowing the decoder to respond to consist commands for the headlight. The new unit, no. 2011 with the SoundTraxx sound decoder, required a few more settings. In addition to controlling the headlight in forward and reverse, I needed to control the bell, horn, and mute functions. This required setting CV21 to 131 and CV22 to 3. (Note: DecoderPro will set all of these CVs for you as you click settings with your mouse).

This combination of settings gave me a matched consist that operates well together with excellent sound and lights. I've provided a table of CVs used for advanced consist operations along with their functions in the chart.

A complete listing of all my final CV settings for both decoders is available on my website (www.dccguy.com), along with a video showing how to speed match locomotives and additional posts on consisting and back-EMF.

CVs for advanced consists

CV number and function
(allowed value)

2* Vstart, starting voltage at first speed step (0 to 255)

3 Acceleration (0 to 255)

4 Deceleration (0 to 255)

5* Vhigh, voltage at maximum speed step (0 to 255)

6* Vmid, voltage at midpoint speed step (0 to 255)

19 Advanced consist address (1 to 127, add 128 for reversed operation in consist)

21 Advanced consist controls for F1-8 (0 to 255)

22 Advanced consist controls for F0f, F0r, F9-12 (0 to 63)

23 Advanced consist acceleration (0 to 255; 0 to 127 = 0 to +127, 128 to 255 = 0 to -127, values are added to CVs 3 or 4 as an offset respectively)

24 Advanced consist deceleration (values calculated the same as 23)

66 Forward trim (0 to 255; 0 to 127 increases speed, 129 to 255 decreases speed)

95 Reverse trim (values for 95 are calculated the same as 66)

245 Econami advanced consist controls, F13-20 (0 to 255)

246 Econami advanced consist controls, F21-28 (0 to 255)

247 Econami advanced consist controls, auto effects (0 to 255)

* Note: CVs 2, 5, 6 comprise a three-step speed curve

4 Turnout control with accessory decoders

1. Using DCC to operate turnouts on your layout opens up several options. Many accessory decoders can be controlled with pushbuttons or toggle switches installed in conventional control panels, like this one on Larry Puckett's Piedmont Southern.

One of the things I miss with my transition from solenoid-based switch machines to slow-motion stall motor switch machines like the Tortoise by Circuitron, SwitchMaster, and others is the ability to use the old, time-tested diode matrix to control and create routes through a series of turnouts.

I always found it convenient to be able to set up a control panel and diode matrix that would allow operators to push one button to align a route up a yard ladder, or through a complex yard. However, the diode matrix doesn't work with slow-motion machines. Consequently, when I switched to Tortoises, I simply wired individual turnouts with reversing toggles.

However, DCC provides additional ways to use accessory decoders to control turnouts and other accessories. This feature allows you to use a throttle to activate turnouts or to use a computer to set up and activate complex routes, turn lights on and off, and basically control almost anything electrical.

When it comes to DCC, many of the old methods for controlling turnouts and other accessories still work. So what options do DCC accessory decoders bring to the mix?

Let's consider how an accessory decoder works. Accessory decoders are similar to mobile decoders in that they are assigned unique addresses. Also, like mobile decoders, commands can be sent specifically to each accessory decoder controlling various accessories.

Some also can be programmed to carry out special functions, such as controlling railroad grade crossing signals, and automating turnout positions, such as in a reverse loop.

Those connected to the DCC power bus can be controlled by a handheld throttle or a computer, and in some cases can provide feedback indicating status to the command station or a computer through the throttle bus.

Many, but not all, accessory decoders can be activated using push buttons or toggles installed on the fascia or on control panels, **1**. Some, like the Digitrax DS44 and Model Rectifier Corp. 1628, can only be controlled using a throttle or computer.

The Digitrax DS64 and NCE Switch-IT Mk2 and Switch8 Mk2 units can be controlled using push buttons and, in some cases, toggle switches. Plus, the DS64 can be programmed to create turnout routes through ladders.

Other accessory decoders may depend on a command station, an optional interface board such as the NCE Mini-Panel, or a computer to create routes. DCC Specialties makes several accessory decoders with programming capability and external controls.

Putting them to work

I use a mix of accessory decoders on my HO Piedmont Southern layout, so let's take a look at some of these units and how to use them. In some locations on the layout I only have one or two turnouts, so fancy programming capability and a complicated control panel aren't required. Instead, a simple clip-on accessory decoder like the DCC Specialties Hare or the NCE Switch-IT Mk2 are good choices, **2**. The Hare clips onto the circuit board of the Tortoise using a card-edge connector.

The Switch-IT is small enough to be attached to a Tortoise using double-sided foam tape or hot glue, and can control a pair of switch machines. Both can be controlled with a throttle or push button switches mounted on the fascia or a control panel.

For sections of the layout with a lot of turnouts in a small area, but not requiring route selection, I like the NCE Switch8 Mk2 and Button Board combination. As the name implies, the Switch8 can control eight switch machines with push buttons, toggle switches, throttles, your DCC command station, or computer commands.

2. The Switch-IT (left) can be attached to a Tortoise with hot glue or double-sided foam tape. The Hare (right) clips directly to the circuit board on a Tortoise switch motor.

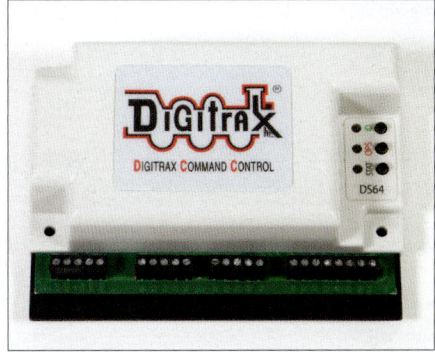

3. Larry attached the Button Board (left) and Switch8 (right) with double-sided foam tape to a piece of hardboard. Then he used a spring-loaded cabinet hinge to install the board under the layout. It flips down when Larry needs to work on it, then flips back up out of the way when he's done.

4. The Digitrax DS64 comes in a plastic enclosure with screw terminals for easy hookup to switch machines, accessories, or push button inputs. It can be programmed to align turnout routes for up to four turnouts. Several can be connected together if they're attached to a Digitrax LocoNet.

I attach the Switch8 and Button Board to a piece of hardboard suspended under the layout with a spring-loaded cabinet hinge, **3**. This allows it to be swung down for wiring and programming, then swung back up out of the way the rest of the time. I then mount the pushbuttons in one or more small control panels attached to the fascia, **1**.

The Switch8 and Button Board can be separated by as much as 6 to 10 feet, and the push buttons can be located as much as 20 feet from the Button Board. However, for runs more than 2 feet, the wires connecting them should be twisted pairs, which you can buy from NCE, or you can twist them yourself.

DCC Specialties offers a series of accessory decoders. The Hare and Tortoise Buddy are capable of controlling individual Tortoises. The dual output Wabbit controls two, and the Jack Wabbit Quad can control and power four turnouts. Among these there are several different versions with specialized control options and functions. The Wabbit, for example, can be programmed to detect a locomotive and line a switch, which is great for automating a reversing loop.

Route control

When it comes to setting up routes, I turn to the Digitrax DS64, **4**.

Each DS64 can control four switch machines using push buttons, input from occupancy detectors, throttles, command stations, or computers. Most importantly, you can program it to simultaneously throw multiple switches to set up routes such as yard ladders.

If you have more than four turnouts in a ladder, or a complex track arrangement, you can use two DS64s and include as many as eight turnouts in a route. Got more than eight turnouts? You can daisy chain even more DS64s and create cascading routes. For routes with more than one DS64, they all must be connected to an active LocoNet, but individual DS64s can be used with any DCC system.

One thing to keep an eye out for is the current rating of the individual control outputs on accessory decoders. Most are rated for about 40mA maximum, which is fine for the Tortoise and most slow-motion switch machines. However, the Micro-Mark Switch Tender pulls about 65mA, which may be enough to let the smoke out of your accessory decoder. Also, most described here won't work with solenoid switch machines (the DS64 is an exception). Other accessory decoders are available that will work with solenoid machines.

Powering accessory decoders is fairly straightforward. All can be powered directly off track power. Just connect them to your DCC power bus and not only will they receive power with which to operate switch motors, they'll also be able to receive commands from throttles, the command station, or a computer.

But there are a couple downsides to attaching them to your main power bus. First, if you have a lot of accessory decoders installed on your layout, the power they consume will take power away from your locomotives, car lights, and any other DCC-powered accessories. Also, if a short at a turnout shuts down your boosters, there will be no power to cycle that switch and clear the short.

Some accessory decoders can be powered using a separate power source, alleviating these problems. Another option is to use a separate booster as a dedicated power supply for all your accessory decoders.

I've given you a lot to consider when it comes to controlling turnouts on your layout, and there are even more choices available that I didn't have space to cover. In addition, many of these accessory decoders are capable of varying degrees of programming. Ben Lake has an excellent series on Model Railroader Video Plus on installing the DS64 and programming routes. You can also visit my website (www.dccguy.com) for more information.

4 Add DCC and sound to a turntable

1. A non-streamlined Virginia & Western class J 4-8-4 steam locomotive takes a spin on the turntable on Douglas Kirkpatrick's HO layout. Douglas added DCC control and sound to this scale 115-foot turntable. *Photos by Douglas Kirkpatrick*

My freelanced HO scale Virginia & Western, featured in the May 2008 *Model Railroader,* is complete—but of course, a model railroad is never really finished. With the major work done, I had time to focus on improving specific aspects of the layout. One of those projects was the turntable at Michaelson Yard. With a bit of work, I was able to turn this direct-current (DC) model into a smooth running, easy-to-index turntable with realistic electric motor sounds, **1**.

Many years ago, I scratchbuilt the scale 115-foot turntable. I animated the turntable using a series of gears for speed reduction powered by a 12V DC motor, **2**, which I reduced to 5V. You can use the same basic techniques to power a commercial turntable.

I used a toggle switch to control the motor and a separate switch to control direction. The turntable ran at a constant speed. Lining up the turntable with an approach track was always a challenge, though, especially for new operators working the engine terminal.

As I was making the switch from DC to DCC, I added sound decoders to my steam locomotives. As I replaced these models' original non-sound units, I found myself with several original, basic 1-amp decoders, one of which I put to use on my turntable.

Before installing the decoder, I first measured the current draw required by the turntable drive motor to ensure that the decoder could handle the load. The motor drew .25 amps, well below the decoder's current rating. I connected the orange and gray wires of the decoder to the turntable motor and the black and red wires to my yard/engine terminal track power, **3**. I temporarily attached the decoder to my programming track to set the basic configuration variables (CVs) including the address, maximum and mid-range speed, and starting voltage.

I set CV2 (starting voltage) to 30. The high value compensates for the low gear ratios inherent in this type of drive mechanism. Without the high setting, the motor turns with the opening of the throttle but the turntable doesn't appear to be moving. I set CV5 (maximum voltage) to 80 and CV6 (mid-voltage) to 60.

As I opened the throttle, the decoder-driven turntable slowly accelerated to maximum speed. Stopping the table at the correct track was easy as the decoder precisely responded to the throttle. This allowed the approach tracks to be placed at any location around the turntable.

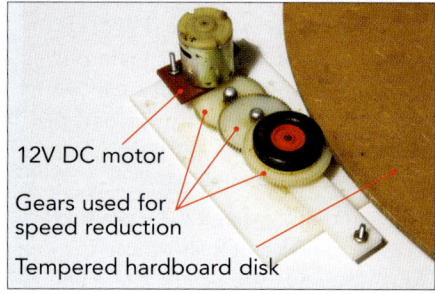

2. When Douglas first built the turntable, he powered it with a 12V DC motor (reduced to 5V). The series of gears was used for speed reduction.

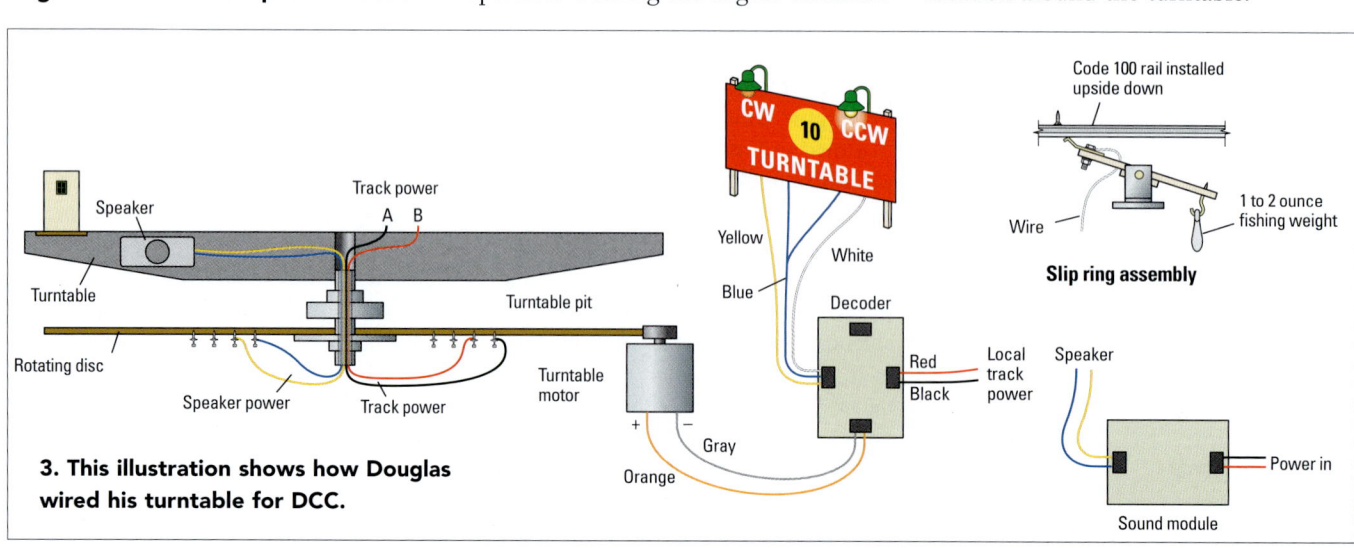

3. This illustration shows how Douglas wired his turntable for DCC.

GO-TO GUIDES
Recommended for You

From the track itself, through the roadbed, and out into the weeds at the edge of the right of way, Jeff Wilson shares the details that keep railroads running, and how to incorporate them into your modeling for a more realistic layout. **Modeler's Guide to the Right of Way** includes 19+ modeling tip callouts throughout the book!

Bring your layout to the next level with a complete overview of modeling railroad scenery in this book from Kathy Millatt. You'll learn how to use the latest techniques and materials, as well as time-tested "classic" methods for modeling mountains, rivers, forests, railroad rights-of-way, towns, and cities that occupy any modeler's layout.

How to Design a Model Railroad will help you transform your basic ideas of what you want in a model railroad into a workable, realistic track plan and overall layout design that you can then build. This is a hands-on guide to drawing and designing a model railroad that will fit the available space and result in a realistic layout.

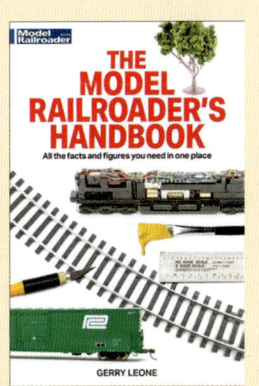

Gerry Leone wrote **The Model Railroader's Handbook** to be the ultimate reference guide, a must-have for modelers of any skill level working in any scale! Get all the facts and formulas you need to know, but can't always remember, in one easy-to-use book.

Kalmbach Media

Shop now at
KalmbachHobbyStore.com

Sales tax and retail delivery fee where applicable.

by Douglas Kirkpatrick

4. This sign, next to the turntable pit, displays the turntable decoder address. The illuminated light indicates the turntable is set to turn counterclockwise.

5. The RailMaster SD1436-8 speaker (inset) conveniently fit between the turntable girders. The speaker faces the control house, and the wires are inside the girders.

6. Douglas drilled a clearance hole through the turntable shaft. He then pulled the wires for the speaker and track power through the opening.

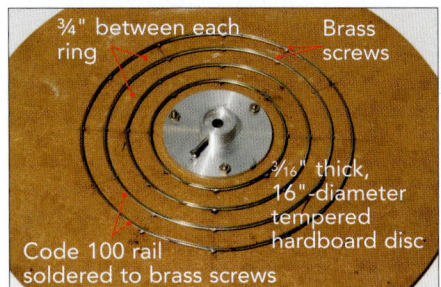

7. Douglas made this electrical slip ring using brass screws and code 100 rail. He soldered the rails to the screws starting in the middle of each rail.

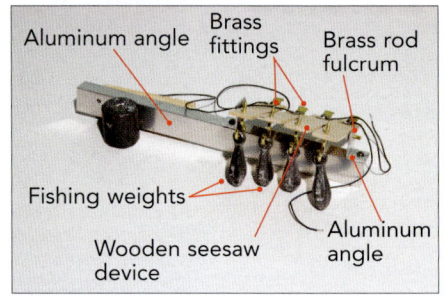

8. This device ensures solid electrical contact between the wipers and rails. The wipers are attached to wood to prevent short circuits.

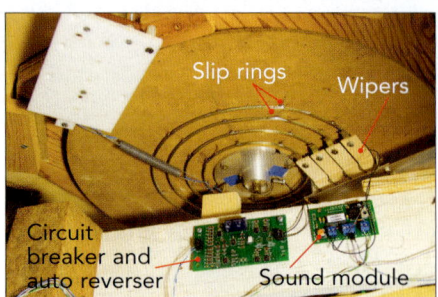

9. Here's what the bottom of the turntable looks like after installation. Power is transferred to the speakers and rails when the turntable rotates.

I built a small billboard sign to display the turntable's decoder address, 4. Two operating gooseneck lights, controlled by the headlight and reverse light outputs on the decoder, illuminate the letters CW or CCW. This identifies whether the table is set to rotate clockwise or counter-clockwise.

Adding sound

I came across an advertisement in *Model Railroader* for ITT Products (www.ittproducts.com). The firm offers many reasonably priced stationary sound modules, from working sawmills to various animal noises. Searching its entire list of sounds, I came across the recording of turntable sound effects.

Instead of mounting a speaker in the turntable pit and trying to hide it, I installed a RailMaster DS1436-8 speaker, 5, between the girders of the turntable at the end with the control house. Now when the turntable rotates, the electric motor sounds rotate with it.

The next step was to provide speaker power through an electrical rotary joint to the stationary sound module mounted under the layout. I drilled a clearance hole through the solid brass turntable shaft to allow for the two speaker wires and two track power wires, 6. In hindsight, I would recommend using a hollow shaft that still would provide both stability and sufficient torque capability. Care must be taken in threading the wire through the shaft hole as well as remembering which wires go to the speaker and which wires go to the track.

On the bottom of the layout, I attached a ³⁄₁₆" thick, 16"-diameter disk of tempered hardboard from the protruding turntable shaft. I used the surface of the disc to fabricate an electrical slip ring. I mounted brass screws in concentric circles around the disk allowing ¾" between each ring. I formed the code 100 rail into tight circles, then soldered the rail to the brass screws starting in the middle of each rail, 7.

Next, I fabricated a set of four contact wipers that slide along each rail. Each wiper is a wooden seesaw device with a brass fitting at one end and fishing weights at the other end, 8.

I passed a rod through the center of each of the contact wipers to form the fulcrum and attached it to thin aluminum angle. The wood ensures that each wiper is insulated from the others. The fishing weights provide the necessary pressure to ensure good electrical contact between the wipers and the rail.

As the turntable rotates, power is constantly transferred to both the speaker and the rails, 9. An auto reverse is required for the turntable track power to eliminate electrical shorts that can occur depending on the turntable alignment with an approach track.

My turntable responds similarly to the prototype I witnessed on the Western Maryland Scenic RR several years ago. Now I can enjoy a slowly rotating turntable, clanging and banging sounds, and the whir of the electric motor on my HO scale railroad.